From Aspirin to Viagra

Stories of the Drugs that Changed the World

Vladimir Marko

From Aspirin to Viagra

Stories of the Drugs that Changed the World

Vladimir Marko
Bratislava, Slovakia

SPRINGER-PRAXIS BOOKS IN POPULAR SCIENCE

English translation by Skrivanek Slovensko, www.skrivanek.sk, of the original Slovak edition published by Ikar, Bratislava, 2018

ISSN 2626-6113 ISSN 2626-6121 (eBook)
Springer Praxis Books
Popular Science
ISBN 978-3-030-44285-9 ISBN 978-3-030-44286-6 (eBook)
https://doi.org/10.1007/978-3-030-44286-6

Project Editor: Michael D. Shayler

This Springer imprint is published by the registered company Springer Nature Switzerland AG
The registered company address is: Gewerbestrasse 11, 6330 Cham, Switzerland

Contents

Acknowledgements

I would like to thank the following experts for so kindly reading parts of the manuscript, providing their expertise and sharing with me their inestimable observations: Dr. Viera Kořínková, CSc; Dr. Katarína Rašlová, CSc; Dr. Katarína Mikušová, PhD; Dr. Danica Caisová; and Dr. Dušan Krkoška, PhD.

To my wife for 44 years of our lives

About the Author

Dr. Vladimir Marko PhD comes from Slovakia. He was born in 1952. He studied at the Slovak Technical University, finishing his studies in organic chemistry in 1975 and his PhD studies in biochemistry in 1980. He then spent ten years working as a researcher at the Institute of Experimental Pharmacology of the Slovak Academy of Sciences before changing his professional life completely by moving into business as a country representative of the Dutch company Chrompack, dealing with analytical instruments. From 1994, he worked for 20 years for the Danish-based pharmaceutical company Lundbeck, first as a representative and later as the Managing Director for Slovakia.

He is the author of numerous scientific papers concerning the synthesis and analysis of drugs and was the editor for a book dealing with drug determinations (Marko V., Ed.: *Determination of Beta-Blockers in Biological Material*, Elsevier Science Publishers, Amsterdam, New York, 1989, ISBN 0-444-87305-8). Dr. Marko has also authored several tens of popular articles that have been published in various Slovak weeklies, dealing with drugs, medicine and history.

He likes running (he has run 20 marathons) and mountaineering, and regards himself as a connoisseur of food and drink.

He has been married for 44 years, with no children.

Preface

Recently, I opened our medicine cabinet at home and found an assortment of 36 different drugs. And honestly, my wife and I consider ourselves to be healthy people. If we look back 100 years, we would have had, at best, bottles of aspirin and quinine. It is very difficult to imagine the world at the dawn of the 20th century, when we were unaware of the existence of very common things used today, like penicillin, insulin or vitamin C, let alone innovations like contraceptives or drugs that fight mental disorders. We did not know a large majority of the drugs that we see as a natural part of our lives today, to the point that often we are not even aware of how we depend on them. Had these drugs not existed, the majority of us would not even be alive today and would not be able to read this book. Our parents or our grandparents would likely have died due to one of the vast number of deadly diseases that have plagued humankind since the beginning of time.

The drugs we know of today have a short history. Until the 19th century, official medicine had no actual need for these drugs. For many centuries, starting from ancient times, diseases were thought to be caused by an imbalance of four basic bodily fluids: blood, phlegm, yellow bile and black bile. Doctors were meant to rebalance these fluids through methods like bloodletting, the use of leeches, serving laxatives, enemas, or substances inducing vomiting. Over the course of one year, the French King Louis XIII received 212 enemas, was induced to vomit 215 times and underwent bloodletting 47 times. His son and heir, Louis XIV, was rumored to have undergone more than 200,000 enemas, sometimes as many as four a day. The French playwright J.B. Molière illustrated the situation with official medicine very well in one of his plays: "*(Doctors) can talk fine Latin, can give a Greek name to every disease, can define and distinguish them; but as to curing these diseases, that's out of the question.*" Those who were reliant on the help of this kind of medicine were more or less out of luck. When George Washington became ill in 1799 and began to complain of neck pain, the medical help that was called upon did everything within their capacity to assist. They induced blisters and let his blood. He ended up losing about two and a half liters of blood, but despite—or more likely because—of this treatment, George Washington died ten hours later.

It was much easier for the common people. They would often seek help from the unofficial medicine practiced mostly by village women. These healers were not interested in the official teachings of bodily fluids. Instead, they would focus more on the objects that they found around themselves. They knew of the properties of many different flowers and herbs and would use them to cure diseases, although this came at the risk of being accused of witchcraft and burned at the stake. Some of these methods are still used today. An example of the difference between official and unofficial medicine was the approach to treating scurvy in the 18th century. Scurvy, as we now know, is a disease caused by the deficiency of vitamin C. Official medicine said that it was caused by a disease of the black bile, which they considered to be dry and cold. For this reason, they considered it necessary to treat it with something warm and moist, such as a broth brewed of barley. They did not use citrus fruits because these were also cold. A Miss Mitchell from Hasfield, located in the province of the Duchy of Gloucestershire, knew nothing of black bile and instead used a mixture of medicinal herbs, wine and orange juice to treat scurvy.

This book contains stories from the history of ten different drugs that have greatly influenced humanity. They are, in alphabetic order: aspirin, chlorpromazine, contraceptive pills, insulin, penicillin, Prozac, quinine, vaccines, Viagra, and vitamin C. The selection of these drugs is often subjective. The author's intent was not to describe the drugs as such, but instead to map the road that was taken to their discovery or invention. The road was often rough, but also adventurous. At the same time, the author looked to record the circumstances associated with their subsequent life. This book is also about the people who chose this path. The majority of them, with the exception of a few charlatans, were inspired by their deep need to help others and by their belief that what they were doing was the right thing, even if by today's standards their methods were harsh.

Vladimir Marko
January 2020

1

Aspirin

No other medicine has such a long and rich history as aspirin. The story of aspirin dates back to ancient times, when people were just beginning to learn of the medicinal properties of willow bark – a kind of predecessor of aspirin. Willow bark was a part of ancient Egyptian, classical, and medieval medicine, though the first scientific mention of its effects comes from the mid-18th century.

It has been over 120 years since aspirin was first introduced as a pharmaceutical preparation, yet even today you would be hard pressed to find a household that does not have those little, white, acetylsalicylic acid-containing tablets in the medicine cabinet. It is hard to imagine that this small wonder can relieve pain, reduce fever and stop a migraine, prevent heart attacks, strokes and thrombosis, and even potentially be used to treat certain types of cancer.

In the stories about the history of aspirin, you will meet a curious vicar, an unrecognized Jewish inventor, and one of the first great captains of industry. You will also meet a spy, uncommonly resourceful marketers, and, of course, plenty of dedicated chemists and physicians. The history of aspirin plays out against the backdrop of the history of the pharmaceutical industry as a whole, allowing us to see how the industry developed from modest beginnings in the late 19th century, when it was something of a pendant on the dye industry chain, all the way through to today, when it is one of the largest and most powerful economic sectors with an annual turnover of nearly a trillion dollars.

The name *aspirin* itself can be contentious. While in some countries, such as Germany, the name Aspirin is a registered trademark with strict rules for capitalizing the first letter, in most countries the word aspirin is used as a genericized name without the initial capital.

V. Marko, *From Aspirin to Viagra*, Springer Praxis Books,
https://doi.org/10.1007/978-3-030-44286-6_1

Story 1.1: The curious reverend and the bark of the willow

Chipping Norton is a smallish, idyllic town in the county of Oxfordshire, England, situated about 30 km northwest of Oxford. According to the most recent census taken in 2011, it has an exact population of 6,337 residents. It is home to the Church of the Virgin Mary which dates back to the 15th century and it also boasts the oldest golf course in the county. The church was recently used as a location for a recording studio, frequented by such legends as Duran Duran, Status Quo and Alison Moyet. We should note that as a village, Chipping Norton is not only idyllic, but also very healthy. Nine out of ten residents claimed to be satisfied with their own health, according to a recent study.

Let us go back 260 years. If a similar study had been carried out in the mid-18th century, the era where our first story takes place, we would have found the results to be the exact opposite. By all accounts, almost everyone would have been afflicted with some of the many common illnesses at the time. High humidity, poor hygiene and malnutrition were, by and large, the main reasons for the majority of diseases. The causes of illness were unknown and treatment methods were from medieval times at best. Essentially, only two medical interventions existed at the time – laxatives and bloodletting – and a visit to the doctor was much more expensive than self-treatment, which only made matters worse. At the time of our story, the Reverend Edward Stone himself was not particularly healthy. When we meet him in 1758, he is 56 years old, quite an advanced age for a man in the 18th century. He lived alone on the outskirts of the town and his position as chaplain in nearby Chipping Norton secured him a comfortable living. Like many older people, Edward suffered with rheumatism.

The rheumatic pain, however, did not prevent him from taking regular strolls beside a small stream that flowed through Chipping Norton, and the town council had planted an alley of willows which made his walks more pleasant. On one such occasion, he unwittingly peeled off a piece of willow bark and put it in his mouth. He was immediately hit with a strong, bitter taste. It reminded him of the taste of cinchona bark, which he was very familiar with as it was the only effective way to treat a fever at that time. That fever was malaria, something the people of the 18th century were not aware of. The good Reverend Stone got the idea to use willow bark in place of the much more expensive cinchona as a way of treating fevers. There was also a scientific reason behind his decision. In the latter half of the 18th century, the prevailing theory was that things which were alike belonged to the same group. Therefore, they believed that each disease should be treated with a substance that in one way or another resembled the disease itself. They used deep blue iris flowers to treat bruises; jaundice was treated with yellow goldenrod flowers; lung illnesses were treated with lungwort leaves, which are lung-shaped. The same principle can be seen in the centuries-long use of crushed rhinoceros horn to increase male potency (the similarity requires no explanation). Sadly, even though many of the medicinal methods that follow this principal have long fallen out of use, rhinoceroses are still being slaughtered in the "Viagra Era."

But let us return to the Reverend Stone's willow bark theory. He reasoned that since fevers arose from the wet and cold, its treatment could be successful using something also found in wet and cold places. This is where the willow comes in and so this willow bark marks the beginning of the history of aspirin.

If you should ever dare to taste a piece of willow bark like Edward Stone, you would instantly be hit with a very intense, bitter taste. Together with this taste, a very small amount of salicin would enter your body. The name salicin comes from the Latin word for willow, *salix*. The salicin that enters the body later decomposes into salicylic acid. This is the exact same salicylic acid created in your body when you swallow that small white pill called aspirin. Aspirin and willow bark have the same active substance that reduces fever, reduces pain and has a whole host of other healing effects. Our distant ancestors similarly experimented with willow bark because their survival depended on their knowledge and use of everything they found around them. They were very likely familiar not only with the distinct taste of willow bark, but also its medicinal properties. Willow bark is most likely one of the oldest natural healing substances in the world.

The first written mention of willow bark, from which we can at least assume the writer was aware of its medicinal effects, dates back 3,600 years. It is on an ancient Egyptian papyrus manuscript called the *Ebers Papyrus* that originates from around 1550 BCE, but some parts of it also contain knowledge recorded from a significantly older period. Measuring 20 m long and 30 cm high, the papyrus contains writing on both sides. It is the most valuable piece of information found about ancient Egyptian medicine and health sciences. It also contains about 700 different magical formulae and medicinal products, 160 of which are plants such as dates, figs, grapes and pomegranates, as well as many different spices and seasonings such as coriander, caraway, fennel, mint and thyme. We can find willow among them, with its effects described in three different places.

Willow bark and willow bark preparations can be found later in history as well. Hippocrates, the most famous doctor of antiquity, also known as the "Father of Medicine," suggested the use of willow bark as a remedy against the pain of childbirth and to reduce fevers as far back as 500 BCE. Around 500 years later, the therapeutic possibilities of willow bark were introduced by Pliny the Elder in his 37-volume work. Another 130 years later, Galen of Pergamon also suggested willow bark to be an effective pain killing method. Given that he was a doctor for gladiators, he had plenty of experience with treating pain.

Willow bark is not the only natural source of salicin. Numerous different plants also contain it, including a plant with a pleasant name, meadowsweet (Latin *Filipendula ulmaria* or, in the older nomenclature, *Spiraea ulmaria*; remember this name, because we will come back to it).

The medicinal use of willow bark and other natural sources containing salicin did not end with the invention of aspirin. Dry willow bark and meadowsweet can be found in today's pharmacies as well, often promoted as a natural alternative to aspirin. The Yupiks are a curious example of how healing practitioners were able to use natural sources as medicine. They are an Eskimo tribe that live in Western Alaska and in Eastern Siberia. They knew of and used the healing properties of *castoreum*, a sticky secretion that beavers (*Castor* in Latin) use to mark their territory. The Yupiks dried the *castoreum* and used it as a pain killer. Modern analytical methods later discovered that *castoreum* has a very high salicin content, as a result of beavers gnawing willow bark to build their dams. *Castoreum* is a valuable resource today, and due to its animal scent it is a sought-after ingredient for perfumes, even by the most popular brands.

Now let us return to the latter half of the 18th century and to Edward Stone, because the story of this English cleric from Chipping Norton is a real milestone in the history of aspirin. His work is one of the first clinical tests of medicinal products in history. It is also an example of how sometimes an incorrect assumption can lead to a positive result.

Given that he scientifically verified his method – that which originates from the moist and cold is treated with that which grows in the moist and cold – he started to collect willow bark. It had to be dried, so Reverend Stone asked the local blacksmith if he could dry it in his workshop. When he dried around half a kilogram of it, he decided to test its effects. He knew nothing of what an effective dose might be, so he started with small amounts. In today's measurement standards, his initial dose was one and a quarter grams of willow bark powder and he served the powder once every four hours. The symptoms started to subside a little bit, so he increased the dose to two and a half grams, causing the fever to disappear completely. We do not know who his first patients were or how he convinced them to take part in his "clinical test." After the first success, word apparently spread around and over the course of five years Reverend Stone treated – and cured – 50 patients of their fevers.

He recorded his entire procedure in a report he sent on April 25, 1763 to the highest scientific authority in England at the time, Lord Macclesfield, President of the Royal Society of London. This report contained a very meticulous description of his procedure, his scientific reasoning and results, which are preserved to this day. His ancient style of writing resembles an extract from a very beautiful work of literature. Considering later developments, the first sentence of his letter is practically a prophecy: "*Among many of the useful discoveries, which this age hath made, there are very few which deserve the attention of the public more than what I am going to lay before your Lordship*."

He was correct in this statement. In most others, however, he had been mistaken. His first mistake was the belief that he had found a way of curing fevers. He had not. He had merely found a way to reduce body temperatures, and it was not due to the fact that willows and fevers originate in moist environments. Today, we know that willow bark contains salicin, from which salicylic acid is created in the organism. We also know that salicylic acid reduces temperatures and relieves pain.

Only a church, an alms-house, three inns and a few other buildings remain from the time of Edward Stone in Chipping Norton, although the alley of willows where the local stream used to flow is still said to exist. Today's willows, however, are just a reminder of the trees that grew there long ago when Edward Stone took his piece of bark to experiment with, his efforts becoming part of the history of aspirin.

Story 1.2: The three fathers and the two miracle drugs

On Monday, August 26, 1856, 18-year-old William Henry Perkin patented his method for producing a purple dye he named *mauveine*. Purple, the color of the powerful, rich and famous, made Perkin rich and famous as well. But the impact of the discovery was far more wide-ranging than just the fame and fortune of one person – Perkin's mauveine ignited the chemical dye industry which would shape the chemical industry for the next 50 years. The dyeing industry would become not only the chief field of the chemical industry but also a very lucrative business and dye factories sprang up like mushrooms in the latter half of the 19th century. Many have long since closed their doors, but some remain in business to this day, one of which is Friedrich Bayer & Comp., incorporated in August 1863.

But let us return to willow bark. A pure form of salicin was isolated in 1829 and it was assumed that this is what gave the bark its healing properties. Surprisingly, salicin is not the substance that reduces fevers and relieves pain, as it is unstable and breaks down in the human body. It is metabolized in part into salicylic acid, which is the substance that actually causes salicin to have antipyretic and analgesic properties – the same effects as willow bark. At the time, it was relatively simple to mass-produce salicylic acid chemically and it was therefore expected to gain prominence rapidly. It did not. There was one major disadvantage to using salicylic acid, in that it caused severe gastrointestinal discomfort. The fact that it is still used today to remove warts is proof of how aggressive the substance can be. Patients who took it complained of severe irritation when swallowing the drug and extreme burning sensations in the stomach. They had to work on finding a compound with two properties – it had to be gentler on the digestive tract and metabolize into effective salicylic acid.

That is where the famous history of Farbenfabriken vormals Friedrich Bayer & Comp. begins (for simplicity, we will refer to it by the company's more familiar name – Bayer). The primary focus of the company was the production of dyes, and they even included that in the name. Bayer was one of the first dye manufacturers to realize that their new-found knowledge could be put to another use, in the development and manufacture of medicines. In 1890, the company established a small pharmaceutical department, staffed by three young scientists who would later become the "fathers" of aspirin: Heinrich Dreser, Arthur Eichengrün and Felix Hoffmann. Each of them had a different personality and each contributed to the development of aspirin in his own way.

Born in 1860, Heinrich Dreser was the oldest of the three and by the time he took up employment at Bayer, he was already a full professor at the University of Bonn. His father was a physics professor, and Heinrich was brought up to be painstaking and methodical. He was a perfectionist and not very popular among his colleagues, but he was highly respected in his field of pharmacological drug testing. His methodological approach is evidenced by the fact that he was one of the first scientists to test drugs on laboratory animals.

Little information remains about his private life, apart from the fact he was a wealthy man. His contract with the parent company contained a clause that entitled him to royalties on the drugs he tested in his laboratory. After he left the company in 1914, he established a pharmacological institute in Dusseldorf. He played the violin, viola, and cello and was married twice; his first wife died and he married his second wife just prior to his own death. He died, childless, of a stroke at the age of 64. There was one more thing he was known for, but we will return to that at the end of this story.

Arthur Eichengrün, the second of the "fathers of aspirin" was extroverted and charismatic, the exact opposite of his older colleague. He was born in 1867 as the son of a Jewish cloth merchant and began his employment in the pharmaceutical department at Bayer on October 1, 1896, where he took over responsibility for the development of new drugs. Unlike the lives of his colleagues, his own life is well-documented and may be worth making into a movie. He was successful, likeable, and rich. He was married three times and had six children. He had what you might call a turbulent life. After leaving Bayer, he went on to establish a successful company that manufactured compounds based on cellulose. The success of the company was built on his inventions (he held 47 patents) and with the start of World War I, military contracts added to that success. When the war ended, he was quickly able to turn his focus from military applications of his products to civil applications.

His company continued to prosper, and it would probably have continued to do so for many years had Arthur Eichengrün not been Jewish. His success came to a halt in the 1930s. With the rise of Nazism in Germany, his company was slowly "Aryanized" and in 1938, ownership was transferred to the Germans. Eichengrün effectively lost all his assets, but his problems were only just beginning. In 1943, he was arrested and transported to the concentration camp in Terezín (Theresienstadt in German). Not only did he survive the concentration camp, but when the war ended he regrouped and continued his scientific work. In spite of the tragedies that struck his life, he lived to a relatively old age, dying in 1949 at the age of 82.

The youngest of the trio, Felix Hoffmann, was born in 1868 as the son of an industrialist. Like his two older colleagues, he received a first-rate education in chemistry. He went to work for Bayer in 1894, shortly after earning his doctorate. He worked in the department that was later taken over by Arthur Eichengrün. Unlike his boss, though, Felix Hoffmann was an introvert and there is even less information remaining about his life than there is about the life of Heinrich Dreser. Once the development of aspirin was finished, he did not continue in his scientific career and instead worked in marketing, where he remained until his retirement in 1928. He died unmarried and without children on February 8, 1946.

In official documents, August 10, 1897 is considered the vital date in the history of aspirin. That was the day Felix Hoffmann prepared acetylsalicylic acid using a process that yielded a quality good enough to make it ready for mass-production and for use as a medicine. Allegedly, one of the reasons for Hoffman's interest in acetylsalicylic acid was purely personal. His father used it to treat his rheumatism but, like all patients, he complained of the terrible taste. A son's love played a role in the history of aspirin.

Felix Hoffmann was not the first to be interested in acetylsalicylic acid. That would be a Frenchman named Charles Frédéric Gerhardt, a professor at the University of Montpellier. He had prepared it nearly 50 years earlier, but while Gerhardt's goal was to study the properties of substances (in his case acids), Felix Hoffmann turned his attention to preparing the compound as the basis for manufacturing a new drug. When you look at how each of these chemists approached the subject, it is easy to see the difference between academic research and corporate research.

When Felix Hoffmann prepared acetylsalicylic acid, his boss, Arthur Eichengrün, immediately sent it to Heinrich Dreser's pharmacology group for research – and he surprisingly refused. He was convinced the new substance would have "enfeebling" effects on the heart. He even said something to the effect of it not having much practical use. He was far more hopeful about another compound he was working on at the time, and like acetylsalicylic acid, it too went down in history. But we will come to that later.

Fortunately, Eichengrün did not give up. He went around Dreser and, after testing the new compound on himself, he secretly sent it for clinical trials to Berlin. The results were good, and when two independent studies subsequently confirmed the benefits of the new substance, Dreser was forced to give his approval, which he did in September 1898.

Thus, Bayer introduced a new drug to the pharmaceutical market. It was registered by the Imperial Patent Office on March 8, 1899 under the name Aspirin. There are three parts to the name: *A* for acetylation, *spir* for the spirea plant, and *in* as the typical suffix for drug names that make them easy to pronounce.

So, who is the real father of aspirin? Right after Aspirin was approved, Heinrich Dreser was tasked with writing about the properties of the new drug for a science magazine. The first scientific information about the new drug was made public in 1899, but there was a catch. He named only himself as the author of the article – maybe out of spite for being forced to approve the drug. His text did not mention Hoffman or Eichengrün, nor give any information about how Aspirin was developed, so the contribution of the other two chemists remained unknown for many years. It was not until 1933 that the first mention of Felix Hoffmann's participation in the development of aspirin was mentioned; Hoffmann is now officially credited as having invented it.

Fifty years later, Arthur Eichengrün wrote a detailed report about the process by which aspirin was invented and, at the same time, claimed to have invented it. He claimed that acetylation of salicylic acid had been his idea and that Hoffmann had only brought his idea to fruition. His arguments were convincing, as was his reason for why his name had never been associated with aspirin. It was inconceivable in the time of Nazism that a Jew would be the inventor of such an important drug. However, Felix Hoffmann is still officially credited for the invention of aspirin, and on its website even Bayer confirms this. We will probably never know the exact role and contribution of each of the three "fathers" in the development of aspirin, but the fact remains that Arthur Eichengrün was the first to have the idea to acetylate salicylic acid, Felix Hoffmann was the first to hold the finished product in his hand, and Heinrich Dreser was the first to describe aspirin.

But more important than who invented it is the fact that aspirin is one of the first drugs that definitely resulted from teamwork.

That could easily be the end of the story about how aspirin was discovered, but there is just one more thing. The acetylation of salicylic acid was not the only reaction of this type pursued by Felix Hoffmann. At the same time, he was trying to acetylate morphine, a substance isolated from opium, which has remarkable pharmacological properties. It effectively relieves pain and acts as a cough suppressant, but it is also addictive, which greatly reduces its possibilities for use. While the acetylation of salicylic acid was meant to prepare a non-irritating alternative, the acetylation of morphine was meant to prepare its non-addictive alternative to morphine. Hoffmann was also successful with this acetylation and although he was not the first in this case either, he was able to prepare a pharmaceutically usable product. The resulting substance, diacetylmorphine, was tested with positive results. It effectively suppressed coughs and was useful for other respiratory problems. The effects were so powerful that some users called it "heroic." This was the substance that Heinrich Dreser had more faith in than aspirin. Bayer released two "miracle" drugs around the same time – one was a drug called Aspirin that reduced temperatures and relieved pain, while the other was a drug that suppressed coughs and provided relief for lung diseases. It was called Heroin. Today, we know that both aspirin and heroin have had a significant influence on humanity, but that is a different history that we will not be covering here. But we do owe you an answer about what was said about Heinrich Dreser. Rumor had it that he was addicted to heroin.

Felix Hoffmann died without any direct heirs, but his two 'children' live on: the immensely useful aspirin and the immensely harmful heroin.

Story 1.3: The industrialist and his business

Carl Duisberg did not go down in history as a famous inventor, but he did go down as a great organizer and leader. Today his name is long forgotten, but during his lifetime it was a name as respected as Rockefeller or Rothschild.

Duisberg was instrumental in the invention of aspirin and in making aspirin the most successful drug in history. He was also instrumental in transforming a small dye manufacturer into one of the giants of the pharmaceutical industry. When he took a job at Bayer as a chemist at the age of 23, it was a small factory that manufactured dyes. When he died at the age of 74, he left behind one of the largest companies in the world.

Duisberg was born on September 29, 1861. His father had hopes that his son would take over the small weaving business founded by Carl's grandfather, but Carl Duisberg wanted to be a chemist. He graduated from high school at the age of 15, from university at 18, and received his doctorate at the age of 20. There was a glut of chemists at the time and it took Duisberg a while to find employment. He signed his first permanent employment contract on September 29, 1884 – his 23rd birthday – with a small dyeworks in Eberfeld, not far from his home town. The dyeworks was called Farbenfabriken vormals Friedrich Bayer & Comp.

His personal development continued along the same trajectory, and six years later he took over the reins of the company. It was at that moment that the company began its transformation from a medium-sized dye manufacturer into a very successful company in an up-and-coming industry – pharmaceuticals. The first step was to set up a small pharmaceutical department with top-notch equipment and hire young, ambitious chemists Heinrich Dreser, Arthur Eichengrün and Felix Hoffmann. We know how the rest of the story goes.

But the invention of aspirin was only just the beginning of its history; it continues today, more than 120 years after Felix Hoffmann first synthesized the white crystals of acetylsalicylic acid.

Carl Duisberg knew from the beginning that aspirin was a potential goldmine for Bayer, but since it was not difficult to produce, there was a risk that other manufacturers would cut into their business. He had to protect his product from the very beginning. Bayer's battle with the competition is a part of the history of aspirin.

Attempts to protect the product proved difficult from the start. The first step was to secure exclusivity by way of a patent, but they had no chance of obtaining a patent in Germany as many chemists before Hoffmann had already synthesized acetylsalicylic acid in various qualities. They focused instead on two large countries – Great Britain and the U.S. – and succeeded. The U.S. patent issued on February 27, 1900 was particularly important for the further development of aspirin and the growth of Bayer. These were only victories in individual battles, however, as the war continued. Competitors in both countries challenged the patents, leading to a period of massive litigation.

Farbenfabriken vormals Friedrich Bayer & Comp. *vs.* Chemische Fabrik von Heyden, one of the largest patent disputes in Great Britain, began on May 2, 1905 and ended 70 days later. At play was the aspirin patent, and the victor's spoils would include not only the entirety of the huge British market, but the markets of the British colonies, including India, Canada, and Australia. Bayer was handed a humiliating defeat.

Duisberg's dreams of a monopoly in the British Isles came to a definitive end on July 8, 1905, but the U.S. market was still up in the air. At the same time the legal proceedings were taking place in Great Britain, similar proceedings were under way in the U.S., against an American company located in Chicago. As in Great Britain, Carl Duisberg and his Bayer were fighting to retain their monopoly on aspirin.

At the time, aspirin sales were gaining momentum in the American market, and in 1905 represented a quarter of all of Bayer's sales in the U.S. In 1909, that figure reached 30 percent. A large part of this success was due to the establishment of the company's own production facility in Rensselaer, New York, in 1903, meaning production did not have to rely on imports from Europe.

There was a darker side to Bayer's success with aspirin, however. Numerous other manufacturers capitalized on their success, and as a result only half of the aspirin sold in the U.S. was the original Bayer product; the remainder consisted of counterfeits. This made the outcome of the American patent lawsuit that much more important for Carl Duisberg. After numerous delays, a verdict finally came down in 1909 and Bayer won.

But the battle with competitors did not end there, as even after the patent was upheld there were still many manufacturers that continued to make aspirin, often of dubious quality. Aspirin was such a profitable commodity that smuggling it in from Europe was worth the risk. Bayer was forced to fight with everything it had, and to promote "their" product they launched a massive advertising campaign – something that was unheard of at the time for ethical drugs. To differentiate their product from the products of competitors, every "real" aspirin tablet manufactured by Bayer was embossed with the Bayer cross logo that intersects at the letter "Y." It was slow progress, but gradually the majority of profits from the aspirin sold in the U.S. filled the coffers of Bayer. Over time, the company that began as a smallish dye manufacturer became the largest chemical manufacturer in Germany.

Carl Duisberg continued to keep a close eye on his child and in 1912, he proudly and ostentatiously opened a new factory on the banks of the river Rhine in Leverkusen. The sprawling complex, taking up nearly 25 acres of land, was one of the largest and most modern manufacturing plants in the world. The complex consisted of dozens of production buildings, huge laboratories, a Japanese garden, and the largest chemical library in the world – all managed with Carl Duisberg's German precision. Leverkusen was the impressive headquarters of the largest chemical company in Germany and Duisberg took up residence in the palatial home situated on the premises. Everything was heading in the right direction, right up until the time Gavrilo Princip fired the shots that led to war, with Germany mobilizing on August 1, 1914.

At first, the situation did not seem all that bad for Bayer, having already lost the large market of one of Germany's foes, Great Britain, long before the war. Their primary market, the United States, was neutral at the start of the war and furthermore, there was a pro-German sentiment among Americans. But then, on April 6, 1917, two years and eight months after the war began, the U.S. declared war on Germany. Suddenly, everything changed.

The U.S. immediately seized control of property belonging to its enemies, including the American subsidiary of Bayer. At that moment, Carl Duisberg lost control over the company's property in the U.S., including the production plant in Rensselaer. But there were many losses still to come. The pro-German sentiment of American society that had been prevalent at the start of the war began to move in a decidedly anti-German direction. Six months after entering the war, the U.S. created the Office of Alien Property Custodian (APC), which was responsible not only for administering the property of the enemy but also for the fate of the property when the war ended. The fate of Bayer's property in the U.S. was definitively sealed in a very short time – just a month after the armistice was signed by the parties at war. On December 12, 1918, the APC announced it would auction off the assets of Bayer in the U.S. All the U.S. assets of the company were offered up in one package that included all tangible property – the production plant in Rensselaer and all the inventory – as well as all intellectual property, including patents and trademarks. One of the most valuable prizes included in the package was the trademark for Aspirin. The property was bought by Sterling Products Inc., for $5.3 million, which today is the equivalent of $123 million. Bayer not only lost all of its U.S. assets, it definitively lost its monopoly on aspirin as well. Carl Duisberg could only stand by and watch.

His activities at home in Germany during World War I are not directly related to the history of aspirin, but should be mentioned anyway. He personally participated in the development of a chemical weapon – the lethal poison gas phosgene – and as a true experimenter, he was one of the first to try out the effects of the gas on himself. He was also active in industrial diplomacy. Following the example of American corporations and cartels, he convinced four of his biggest competitors – BASF, Hoechst, Agfa (or rather, its predecessor), and the now little-known Leopold Cassella & Comp. – to create the German coal tar dye industry syndicate (*Interessengemeinschaft der deutschen Teerfarbenfabriken*). From the beginning, this giant was known as IG Farben; our story will come back to this syndicate.

No matter how cruel December 12, 1918 may have been for Carl Duisberg and his aspirin, life went on and so did the history of aspirin. Aspirin was put to a first test soon after the end of World War I. From 1918 to 1919, a massive flu pandemic raced around the globe in several waves. It went down in history as the Spanish flu and remains one of the greatest pandemics in human history. Conservative estimates put the death toll at around 18 million people, the same as the number of casualties in the war that had just ended. Aspirin was the only effective drug at the time. Although it did not cure the flu, its ability to reduce fever gave many patients a chance to recover. The role of aspirin in the pandemic was later questioned, but the majority view remains – without aspirin, the number of casualties would have been significantly higher.

It was actually this pandemic that kicked off the golden age of aspirin. Nobody had a monopoly any longer on the production or the name, and over a short period of time there was a veritable explosion in the number of aspirin manufacturers around the world, but primarily in North America. The "real" Aspirin – the one bought at auction by Sterling – was suddenly faced with hundreds of competitors. Manufacturers offered aspirin under various names and in various combinations. Along with Aspirin, pharmacies stocked Aspro; Calaspirin (a combination of aspirin and calcium); Cafiaspirina and Anacin (aspirin and caffeine); Alka-Seltzer (aspirin with sodium bicarbonate); and many others. These versions of aspirin were practically indistinguishable from one another, and only the advertising campaigns of their manufacturers and the amount each manufacturer was willing to invest in marketing could make or break any of them. This created enormous pressure, which resulted in aspirin gradually becoming a mandatory component of every first-aid kit, and it could be found in practically every woman's handbag. The little white tablet for pain and fever became an icon of interwar America. The period from 1919 to 1941 is commonly referred to as the "aspirin age."

Carl Duisberg was still around to witness many of these events but, fortunately for him, he passed away before the fate of IG Farben became known. He was not there to see how "his" syndicate became one of the biggest supporters of the Hitler regime. The syndicate ceased to exist when World War II ended, and 23 of Duisberg's former colleagues stood trial before a military tribunal at Nuremberg. Ten of them were convicted of war crimes.

Carl Duisberg died on March 19, 1935, but two years before his death he personally illuminated the giant Bayer logo in Leverkusen. It was the typical Bayer cross placed between two factory smokestacks. At 236 feet (72 m) tall and illuminated with 22,000 light bulbs, it was the world's largest illuminated advertisement at that time.

By the time Duisberg died, he had received many honors. Streets and squares in several German cities were named after him, and an education institution still bears his name. He was a visionary who brought the pharmaceutical industry full circle – from a mere appendage of the chemical industry to its own separate industry, and one of the most profitable. According to the obituaries, he was "*a man who may be regarded as the greatest industrialist the world has yet had.*"

Story 1.4: The great German patriot and the Great Phenol Plot

As mentioned, the United States remained neutral at the beginning of World War I and there was a prevailing pro-German attitude. This is not surprising. Great Britain stood on the other side of the front and the memory of British rule – that had ended 130 years before – was still etched in the collective memory of the American people. The fact that Great Britain, with the most powerful navy in the world, blocked trade between the central powers (Germany, Austria-Hungary, and the Ottoman Empire) and the rest of the world, including the U.S., did nothing to ease the anti-British sentiment. The U.S. was beginning to feel the shortage of basic raw materials, one of which was phenol. This is a simple organic compound obtained primarily from coal tar, and is a precursor to many more complicated compounds, of which aspirin is one. It is also used to produce trinitrophenol, a substance similar to trinitrotoluene (TNT), which is also used to manufacture explosives. Since it was wartime, phenol was naturally an important strategic material and Great Britain was not only preventing the enemy from trading phenol, it was carefully guarding phenol for the needs of its own military.

The Bayer factory in Rensselaer only handled the final phase of aspirin preparation, the acetylation of salicylic acid, and the company purchased the input raw material from its American suppliers. However, if there was no phenol, there were no input raw materials, and if there were no input raw materials, there was no acetylsalicylic acid. The situation was so critical that the factory almost stopped production in 1915. Fortunately for Bayer, and for aspirin, there was another American who needed phenol. This was Thomas Alva Edison, the wizard of Menlo Park, considered the most prolific inventor in history and the holder of an incredible 1,093 patents. He needed phenol to produce condensite, which is a resin chemically similar to Bakelite. Edison used it to make discs for one of his most famous inventions, the phonograph. Like Bayer, he was also facing a phenol shortage, so he solved the problem by opening his own factory to produce phenol. By June 1915, he was producing 12 tons of phenol per day. Since he only needed nine tons for himself, three were left over. This is where Dr. Hugo Schweitzer steps into the history of aspirin.

Dr. Hugo Schweitzer was German. He was born in 1861 in Upper Silesia, which at the time belonged to Prussia and is now a part of Poland's Warmia-Mazury Province. His contemporaries remembered him as a friendly, energetic, and goal-oriented young man. These traits would characterize him until his untimely death.

After completing his chemistry studies, he briefly worked in Ludwigshafen, Germany in the research laboratories of *Badische Anilin und Soda Fabrik* (later known as BASF). He did not remain long in research, or in his homeland, emigrating to the U.S. in 1889 and settling in New York. He obtained U.S. citizenship in 1894.

Schweitzer's remarkable and controversial career began in New York. After arriving there, he quickly became part of professional society and was soon a prominent member of New York's chemical community. He is no doubt the only American chemist who can boast of being both secretary of the American section of the British Society of Chemical Industry and the chairman of the American chapter of the German Chemical Society (*Verein Deutscher Chemiker*). In 1904, he became one of the first presidents of the prestigious Chemist's Club in New York. He was highly regarded as an excellent organizer and passionate orator. When William Henry Perkin – the very same Perkin who set chemical production in motion at the age of 18 – arrived in the U.S. in 1906, it was Schweitzer who delivered the welcome speech in his honor.

However, Dr. Hugo Schweitzer was first and foremost a German with a strong belief in the cause of his homeland. Soon after World War I erupted, he was deeply involved in pro-German propaganda. He was an experienced organizer, wrote regularly for a weekly periodical called *The Fatherland*, and founded a company in the U.S. that published translations of the great German authors. He himself translated a German songbook into English. He was a member of the German propaganda council in the U.S. and was even its chairman shortly before his death. Primarily, though, he knew how to use his standing and contacts and he published and lectured unflaggingly to promote the German Empire.

He crossed paths with Bayer in 1897 when he accepted employment as head of the company's pharmaceutical laboratory in New York. He was companion and chief adviser to managing director Carl Duisberg on his journey to the U.S. in 1903, earning his trust to such an extent that Duisberg named him president of the *Synthetic Patent Company* which managed Bayer's U.S. patents. Although he would later strike out on his own, Schweitzer remained in contact with Bayer – which is how he learned of the problems with phenol.

He knew that Bayer was suffering a shortage of this raw material, and he also knew that Edison's factory had begun producing three tons more every day than

Edison required. Schweitzer did three things. He signed a contract with Edison to buy all of his excess phenol from July 1, 1915 through March 1916, at a price that was substantially higher than the market price. He also made a deal with the manufacturer of the intermediate product needed to produce aspirin, whereby the manufacturer would immediately buy the phenol that would be used by Bayer to make salicylic acid. Schweitzer could use the remaining phenol that would not be used to manufacture aspirin any way he wanted. He then established a company called the *Chemical Exchange Association* through which the deal would be organized.

It all seemed legal and sound, and everyone was happy – Edison, Bayer, and Schweitzer – until the day of July 24, 1915, less than a month after the deal had been set up, when Heinrich Albert lost his briefcase on the El train heading toward Upper Manhattan; and who should discover it but none other than the American Secret Service agent who had been following Mr. Albert.

Heinrich F. Albert was not just anyone. He had official duties as the commercial attaché at the German embassy in the U.S., but he was responsible for "other" pro-German activities in America, from propaganda to espionage. More than $30 million passed through his hands, an exorbitant amount in today's money. His activities did not escape the notice of the Secret Service, which assigned agents to follow him – and ultimately the lost German briefcase ended up in the hands of the Americans.

The documents found in Heinrich Albert's briefcase indicated that the money used to finance Dr. Hugo Schweitzer's phenol deal with Thomas Alva Edison had come from the covert sources of Heinrich Albert. An ordinary business transaction suddenly turned into an anti-American conspiracy. According to a calculation that was later made public, the contract he signed with Edison had allowed Schweitzer (and therefore, Germany) to swindle the U.S. out of 680,000 tons of phenol that could have been used to produce over two million tons of explosives, enough to fill four freight trains of 40 cars each. While that calculation is a slight overestimation, it truly was a massive amount.

There were ultimately no legal repercussions, as the documents did not prove to be incriminating enough to bring charges against any of the co-conspirators. In addition, the U.S. was still a neutral country at the time so it was really just "business as usual," although very advantageous business if you were the German embassy. The entire conspiracy ended with some articles in the anti-German newspaper, the *New York World*, and a few columns in *The New York Times*. While Heinrich Albert hid out at the embassy, Hugo Schweitzer defended himself by claiming that his actions had saved the lives of many soldiers on the front, Bayer had obtained sufficient salicylic acid, and Thomas Alva Edison had sold the rest of his phenol to the U.S. military. When Dr. Hugo Schweitzer died of pneumonia just before Christmas in 1917, the obituary in *The New York Times* only noted his pro-German position and his propaganda activities.

The great phenol conspiracy made a comeback in mid-1918, when the activities of the Office of Alien Property Custodian went into full swing under the leadership of A. Mitchell Palmer, an attorney, unsuccessful candidate for senate, and ambitious public official. The office became one of the largest businesses in the U.S., unsurprisingly, as it gradually seized and administered all the property of German companies situated in the United States. That property was worth nearly $1.5 billion (U.S.), which would be $30 billion in today's money. Palmer worked closely with the special Bureau of Investigation, an earlier name for the FBI. In mid-1918, the chief of the Bureau of Investigation raised the issue of a German conspiracy against the interests of the U.S. The details were no different from those that were known in 1915, but the actors had changed. In 1915, the Great Phenol Plot was the anti-American conspiracy of the German embassy in the United States. Money poured through the embassy's commercial attaché and the entire deal was conducted by a German propagandist. In 1918, Bayer was added to the scheme. In its revelations, the Bureau of Investigation named Dr. Hugo Schweitzer as the director of the American subsidiary of Bayer and with this false claim, Bayer officially became a co-conspirator, despite the fact that the company had used only a small portion of all the phenol the conspirators bought from Thomas Edison. Along with being tagged the director of Bayer, Hugo Schweitzer was also named as German spy number 963192637. The number was allegedly assigned to him by the Imperial Ministry of War, a ministry that never actually existed.

But no one spoke up. Dr. Hugo Schweitzer was dead, Dr. Heinrich F. Albert was pursuing a political career in Germany, and the American subsidiary of Bayer was fully focused on trying to save its assets – which we know they failed to do. Six months after the company was implicated in the great phenol conspiracy, A. Mitchell Palmer sold all the American assets of Bayer in the course of one afternoon.

Story 1.5: The man from New Zealand and marketing magic

World War I ended in 1918, and with it the aspirin monopoly of Bayer. The company had long before lost the patent in Great Britain and across the Commonwealth. The U.S. patent expired in 1917, and when its assets were sold at auction in December 1918, Bayer also lost its trademark in the United States. Practically every producer could now use the name *aspirin*. When World War I ended, a new war could begin – the aspirin wars. In just a few years, dozens of manufacturers cropped up in the U.S. alone, each of them offering the same thing – acetylsalicylic acid pressed into small white pills. Every now and then, one of them would add a miniscule amount of something else, like caffeine or calcium, but they made no difference to the effectiveness of the drug. Marketing was what was needed.

Success depended on each manufacturer's cleverness, and on how much money they were willing to part with. Just as soon as a new medium came on the scene, aspirin manufacturers added it to their marketing mix. Newspapers and magazines were a given, and when radio was invented and became more widespread, aspirin commercials were aired during prime broadcasting hours, much to the delight of radio station owners. With the rise of automobiles, billboards began appearing on roadsides and many of them hosted aspirin advertisements. Even automobiles themselves were an advertising medium.

There were many marketing magicians in the history of aspirin. One was Max Wojahn, a German emigrant and chief of the export department at Sterling for Latin America, the same Sterling that bought Bayer's assets at auction. When the company entered the Latin American pharmaceutical market in the 1920s, the market was completely unregulated. Pharmacies were brimming with aspirin (and other drugs) of various provenance and quality. Apart from the aspirin manufactured by Sterling in the U.S. (with the Bayer logo stamped on the tablet), there was locally-produced aspirin and aspirin smuggled in from various parts of the world, including the "original" aspirin manufactured in Germany also bearing the Bayer logo on every tablet.

Max Wojahn concentrated on the most difficult product to replicate – Cafiaspirina, aspirin with a bit of caffeine added – and he began to promote it. In short order, Cafiaspirina advertising was everywhere; in newspapers, on posters and billboards, on cars, newsreels, and later radio. He created a traveling sales team that was the largest and most effective on the entire continent at the time, with massive advertising costs to match. In the late 1920s, promotional costs represented 15 percent of sales; by 1934, it was 55 percent Sales grew at the same pace, however, and in 1929 the net profit from aspirin sales in Latin America was $1.25 million (the equivalent of $17 million today). But even that was not enough for Max Wojahn. To spread the news about Cafiaspirina to the most remote areas, he came up with an entirely new form of promotion. He turned his sales force into roving movie theaters, equipping their vehicles with gramophones, movie projectors and an electric generator, and sent them off to sell the product. If they were unable to reach a remote area by car, he put the mobile movie theaters on boats. The procedure was always the same: The sales rep would show up, wait for it to get dark, set up the generator, then turn on the projector and start the show. They showed various programs, such as old American newsreels, Mickey Mouse cartoons, short silent movies, and the like. The content was not really the point, as most members of the audience were seeing a movie for the very first time. Courtesy of Max Wojahn and his Cafiaspirina, the indigenous peoples living high in the mountains were inducted into the age of mass media.

Of course, that was just a side effect of the entire activity, as the most important part of the show was the aspirin commercials inserted between movie clips.

Because of him, the natives in the remotest of places in Latin America learned about the white pill that could ease headache pain and treat the flu. Even though that second claim was not true, we can imagine that the drug must have seemed like a miracle that could get rid of a horrible headache.

For radio advertising, they used jingles – short musical slogans – promoting aspirin. Rising star Eva Maria Duarte had a small singing part on one of the jingles. When she once failed to show up for a recording session she was promptly fired, but the firing was immediately reversed with deep apologies because Eva Maria was the girlfriend of dictator Juan Perón. She is remembered by history as his wife – Evita. Maybe it was in part because of Evita that Argentina became the world's biggest consumer of aspirin per capita.

But Max Wojahn was not the uncrowned king of aspirin, nor were any of his colleagues in South America. That accolade belongs to Herman George Tankersley Davies, better known as George Davies, a striking and inimitable character. A thunderous, always optimistic man, he always ordered two beers so he would not have to wait for the second one. He was fortunate to be born in an era that looked favorably on enterprising, creative people. In his day, there was no such thing as political correctness, ethical marketing, or privacy protection.

He was born in New Zealand in 1882, orphaned as a child and left to fend for himself. His businesses failed, one after another, and with creditors on his heels he absconded to Australia where he took a job as a traveling salesman for a printing company. One rainy December afternoon in 1917, he knocked on the door of G. R. Nicholas & Co.

This company was owned by the brothers George and Alfred Nicholas, the first of whom, George Richard Rich Nicholas, was a pharmacist. In 1915, after multiple unsuccessful trials during which his simple laboratory nearly burned down, he made aspirin. The quality was good enough to obtain a license from the Australian authorities to manufacture and sell it, but he did not have the cash needed to start up production, so he recruited the help of his older brother, merchant Alfred Michael Nicholas. They named their product Aspro to distinguish it from the generally known aspirin. It was a sort of code name, a portmanteau of the words NicholAS and PROduct. Aspro remains the second most-used commercial name for aspirin.

Aspro was not very successful in the beginning, and sales did not even cover the costs of the input chemicals or the simple equipment needed to make the aspirin. G. R. Nicholas & Co. was burdened with heavy debts and would likely have gone under had George Davies not come knocking at the door. It did not take long for him to sell the brothers on his marketing plan. He first selected a trial area, the Australian state of Queensland, and asked them to produce some samples for him, worth three cents per pack, to be handed out. This would also require a massive advertising campaign, so he asked for a one-week advance on his salary of four pounds and a one-percent share of the sales.

The Nicholas brothers' company provided the Davies campaign with around 150,000 packs of Aspro, worth 2,000 Australian pounds, the currency at the time. It was not the first time in history that a marketing campaign had used such a large quantity of samples, but it was the first time such a quantity was used to promote a reputable drug.

The strategy was successful and the results triumphant. As predicted by George Davies, when people tried the samples and found out that Aspro really worked, they were willing to buy the product. Aspro sales skyrocketed in Queensland as well as in the other Australian states. And not only there. In 1924, the Nicholas brothers' company, now called Nicholas Proprietary Ltd., successfully expanded to the market in Great Britain. They followed the same plan – handing out samples, backed by a huge marketing campaign.

In our time of advanced political culture, the advertising style of George Davies's slogans seems inconceivable. He liked to name-drop and the information he proffered was frequently very distorted. He found inspiration everywhere. Just take one example:

"Hats off to Abe Lincoln, because he was always a stickler for the truth. And as such, he would recognize a wonderful invention like Aspro in a minute."

Are you thinking that makes no sense? That did not bother George Davies. He did not hesitate to use made-up stories with a melodramatic subtext. In one of these stories, the protagonist was a detective who apprehended a perpetrator because Aspro helped him keep a clear head at the critical moment. In another, a nurse was freed by Aspro after 15 years of agony. The real gem was a thank you letter from one Able Seaman Jevons, who served in a submarine. The trying ordeal of serving at the bottom of the sea gave him unimaginable headaches and the only thing that helped him was Aspro tablets. A picture of the submarine and the letter writer were placed next to a copy of the letter. It was all made up.

George Davies was one of the first people to understand that the way to sell any product is to make it interesting. Aspirin was not very interesting; it was pure, safe, had a predictable effect, and it could be trusted. But if it was to be successful, it needed a touch of the extraordinary, and that's what George Davies was all about. He could come up with advertising slogans whenever and wherever he was; in rented office space, in snack bars at the train station, in houses of ill repute. His efforts would have all been for naught had the underlying product not been a good one, however, and among the aspirin brands available at the time in England, Aspro was one of the best.

But times were changing, and George Davies did not. He could not adapt, and soon began to lose his influence over the company's marketing. After the end of World War II, he decided to turn his attention to finding a reliable system for winning at roulette, even installing a roulette wheel in his office. He died a wealthy man surrounded by beautiful women; after all, he still had his one percent of royalties from Aspro sales.

George Davies belongs to the past. His work methods would most certainly crack under the weight of today's laws and regulations, but in the words of his biographer, he was "*...one of the people I would have liked to have met.*"

Story 1.6: The country doctor and medicinal gum

During World War II, military procurement was the source of a great boom in the chemical industry (among others). Once the war ended the military orders dried up, so alternative peaceful uses were sought for existing production capacities. Many manufacturers chose drug manufacturing as a suitable option, and the extremely successful aspirin, which was relatively simple to produce, was a popular choice. After the war ended, the number of manufacturers and brands of acetylsalicylic acid, pressed into white pill form, would soon return to pre-war numbers. In the U.S., the Bayer brand of aspirin, manufactured by Sterling, faced a total of *220* rival aspirin brands. One way some manufacturers tried to stand out from the crowd was to add a small amount of other substances to the aspirin. These added substances were of very little practical use, but they could be used to market the product as being "different" from aspirin, i.e. better. The first to show up was Anacin, a combination of aspirin, acetanilide, caffeine, and quinine. Then came Bufferin, which was aspirin with two antacid substances. The original idea, and it was not a futile one, was to use Bufferin to treat hangovers. The aspirin was for the headache, and the antacids were to neutralize stomach acids. Excedrin was introduced in the early 1960s, containing aspirin with caffeine and two other substances. All of the big manufacturers laid out huge amounts of money for marketing. In the 1950s, the manufacturer of Anacin was investing around $15 million annually and the manufacturer of Bufferin only lagged behind by two million. It is not surprising that aspirin sales, and therefore consumption, continued to rise. It was available over the counter, the price was reasonable, and it became a permanent part of the lives of Americans.

The situation was similar in Great Britain, where the biggest players were Aspro (with George Davies behind the scene); Anadin, which was the British version of the American Anacin; and Disprin, a soluble version of aspirin. They each wanted to corner the aspirin market and they all sank substantial amounts of money into the promotion of their product. The global consumption of aspirin was so high that in 1950 – a year after aspirin turned 50 years old – it was entered in the Guinness Book of Records as the most frequently sold pain reliever. In the early 1950s, everything seemed to be coming up roses for aspirin.

But for the really big players, the success of aspirin was accompanied by huge costs and growing competition. Some of them began looking for other molecules with properties that would rival aspirin. In a twist of fate, the company that began taking bites out of the aspirin pie had the same name as the company that was there at its inception: it was the British company Bayer Ltd.

Despite its name, Bayer Ltd. did not belong to the parent company Bayer that was headquartered in Leverkusen. It was part of the American concern Sterling, the very same concern that bought all the property of the American subsidiary of Bayer at auction after the end of World War I. Regardless of ownership, Aspirin was not thriving in Great Britain. Aspro and Disprin had practically cornered the market, which was detrimental to Aspirin sales. A new product was needed for survival.

Although aspirin was the dominant product among pain relievers, it was not the only substance to have those properties. Part of the success of aspirin was in the fact that molecules to rival aspirin remained forgotten for the most part. One of these was a substance known as acetaminophen. Its chemical name is N-acetyl-para-aminophenol and it would later be given the generic name paracetamol, which is still used today. Acetaminophen was first prepared in 1878, earlier than aspirin. When it was rediscovered and clinically tested, it was found to have the same pain-relieving qualities as aspirin and was without significant side effects, which meant it could be given in higher doses. In 1956, Bayer Ltd. introduced a new pain reliever in Great Britain called Panadol.

Where acetaminophen/paracetamol surpassed aspirin was in gastrointestinal complications, especially stomach bleeding. While aspirin can cause an upset stomach – even though very few people experience this side effect – and can also cause light bleeding, paracetamol does not have these side effects. While bleeding after swallowing aspirin is negligible, when the clever marketers multiplied that by the population of Great Britain, they arrived at a number that would fill several swimming pools. The result was an advertising metaphor that even George Davies would be proud of: swimming pools filled with blood. The advertising campaign was so effective that whenever someone wanted to buy aspirin from a pharmacy in 1950s Great Britain, the pharmacist would usually ask whether the customer was aware that aspirin would put a hole in their stomach.

Despite the efforts of the manufacturer, however, Panadol sales grew at a snail's pace. Aspirin was aspirin, after all. In 1967, ten years after paracetamol was introduced, it still had a usage that was three times lower than that of aspirin; however, nine years later, in 1976, paracetamol consumption had nearly caught up with aspirin consumption.

Paracetamol was approved for adult use in the U.S. under the name Tylenol in 1960, a bit later than in Great Britain. The trend in consumption followed the same curve as in Great Britain, and in 1976 Tylenol consumption had exceeded the consumption of the best-selling aspirin.

Another rival had already entered the market, however: ibuprofen, which to this day is sold under the name Brufen. From that moment on, the analgesic (pain reliever) market was split into thirds and the importance of aspirin began to decline. Large quantities continued to be sold and consumed, but the heyday of aspirin had ended. Once the British pharmacologist John Robert Vane and his colleagues first described the mechanism of action of aspirin in 1971, aspirin would not recover its lost ground. By the mid-1980s, the share of aspirin sales had fallen to six percent in its very own bastion, the United States. Aspirin was on the way to being forgotten.

At the same time, it had long been known that aspirin had properties that made it useful for more than just fever reduction and pain relief; it also reduced blood clotting. That aspirin increases bleeding had been known practically since it came into existence. It was considered an adverse effect and that is how it was viewed. The first to point out how to use the "adverse effect" of aspirin to benefit patients was the American doctor Lawrence L. Craven, who wrote his observations regarding its use in cardiovascular diseases in 1950. Unfortunately for patients, Dr. Craven was not a renowned pharmacologist or epidemiologist. He was "just" a general practitioner from Glendale, California.

Dr. Craven came across as a bit old-worldly. He was tall, always well-groomed, wore glasses with a metal frame and an unobtrusive bow tie rather than a regular tie. He specialized in ear, nose and throat medicine and routinely performed tonsillectomies. He prescribed Aspergum, a chewing gum containing aspirin, for his patients for pain relief after surgery. The typical dosage was four pieces of gum per day. He began to note that post-operative bleeding in some patients took an unusually long time to stop and, in some cases, never did, with the patient having to be hospitalized. Upon further observation, he discovered that the patients with increased bleeding were experiencing more severe pain and instead of the prescribed four pieces of chewing gum, they were using much more. Some were chewing as many as 20 pieces, which was the equivalent of 14 aspirin tablets. Dr. Craven was not only an astute observer, he also knew how to make the connections and he reasoned that aspirin prevented thrombus (blood clot) formation. It was this that was preventing blood from clotting after a tonsillectomy. If it was preventing the formation of thrombi after a tonsillectomy, it could also prevent their formation in the blood vessels – known as thrombosis. Since thrombosis is a frequent cause of myocardial infarction, it could be assumed that aspirin would be effective in the treatment of this often-fatal disease.

It was a bold hypothesis, but Dr. Craven set out to verify it. He selected 400 men, to whom he prescribed 2–6 daily aspirin tablets. Not one of them had cardiovascular issues. Over time, he gradually increased the number of patients; in 1953 there were 1,500 and three years later there were 8,000. Not one patient suffered a heart attack. Dr. Craven was able to conclude his research by saying that, "*the administration of aspirin is a safe and reliable method of thrombosis prevention.*"

The question that remains unanswered is: Why did this astonishing discovery not receive a deserved response from the medical community? The simple answer is that Dr. Craven did not have the clout of a renowned expert; he was just a suburban doctor and his work did not meet the requirements of serious clinical trials. Not only that, he reported his findings in obscure medical publications that had low readership. It is even possible that he embellished his results a bit – not one case of a heart attack in such a large group is borderline improbable. However, none of those things excuse the fact that ignoring his work, and ignoring aspirin's preventative effects in myocardial infarction treatment, cost the lives of millions of people.

From the time of Dr. Craven's discovery that aspirin was effective against cardiovascular diseases, there would be many more twists and turns along the way to official recognition. Various large clinical studies were unsuccessful, some had to be interrupted, and the results of others had to be revised. Nearly 40 years would pass from the time this general practitioner from California reported his findings until his results were officially verified and the administration of aspirin to prevent thrombosis was officially approved. Finally, in the late 1980s, aspirin went from being a simple analgesic and antipyretic to a drug that could prevent a fatal heart disease – a heart attack. Aspirin had received a new lease on life.

The prevention of a serious disease like myocardial infarction, as a heart attack is technically known, was only the beginning of aspirin's new life. The formation of thrombi in the blood vessels causes other serious illnesses, one of which is sudden cerebrovascular accident – a stroke. Aspirin prevents the formation of thrombi, so preventive administration reduces the likelihood of this illness.

But the surprising discoveries do not end there. Studies on thrombus prevention over the past decades, coupled with the identification of aspirin's mechanism of action, have literally opened the floodgates to a massive flow of clinical studies in many other fields of medicine. The result has been that today, aspirin can have a positive effect on 50 different symptoms, can be used to treat 11 conditions, and can be administered in the prevention of 12 others. Apart from the classic illnesses that aspirin has been treating for over 100 years, we can add migraines, joint conditions like arthritis, rheumatoid arthritis, psoriatic arthritis, and rheumatic fever. Then we have blood clotting disorders, fibrillation, coronary artery disease, Alzheimer's, diabetes complications such as retinopathy and nephropathy, and others.

Unfortunately, Dr. Lawrence Craven was no longer around when any of this came about. He died in 1957, one year after publishing his last paper, and before the first recognized scientific study that would confirm his findings. He spent his life trying to prove the effectiveness of aspirin in preventing heart attacks. Ironically, he died of a myocardial infarction.

Concluding remarks

Aspirin began its existence as a simple, effective, and well-tolerated pain reliever and fever reducer. Ninety years later, it continues as a simple, effective, and well-tolerated drug that prevents heart attacks and strokes. And it seems we have not heard the last of aspirin.

About 25 years ago, before the end of the last century, the first information began to appear that long-term use of aspirin may lead to a reduced risk of cancer (in animals only at that point). Shortly thereafter, retrospective observational studies began of cancer incidence in long-term users of aspirin. The results of these studies are more than encouraging, but for now the use of aspirin for cancer prevention has not been officially approved. However, chances are that new indications will soon come down the pike. It is no wonder that aspirin has lately been labeled "the wonder drug".

Aspirin has led a rich life, like no other drug in the world. It has been entered in the Guinness Book of Records twice; first, in 1950 as the most frequently sold analgesic, and then in 1999 when, as part of the celebration of its 100th anniversary, the Leverkusen headquarters building was transformed into the world's largest aspirin package. They needed over 26,000 square yards (21,700 square meters) of fabric for this feat.

In 1952, aspirin climbed Mount Everest in the company of Sir Edmund Hillary. It climbed even higher when it was included in the first aid kits of the spacecraft on the Apollo missions: Apollo 11 through Apollo 17. It was used as a stable currency during a time of hyperinflation in South America in the last century when money was fast becoming worthless. In 1997, a beautiful, pale pink rose was cultivated and named in its honor.

Today, aspirin is the most frequently used drug in the world. It is estimated that around 120 billion aspirin tablets are consumed annually – that works out at 44,000 tons to the tune of about $2 billion. A total of 1,500 scientific papers on aspirin are published annually, although some authors have said the number is 3,500.

Most importantly, even 120 years after it was discovered, aspirin is one of society's most essential drugs. It is safe, cheap, and so widespread that we cannot even imagine what we would do without it. The wonder drug.

2

Quinine

Malaria is something you only encounter today when you travel to tropical regions like central Africa, some parts of South America, and the Middle East. It is a disease we associate with exotica and adventure. Even in the history books and literature, it is usually mentioned in the context of discovering unknown lands. It is difficult to imagine that malaria was once a common disease in Europe. We encounter it in the very first story as a serious problem in Roman history, and even the English Reverend Edward Stone from the willow bark story was familiar with it.

Briefly, malaria is an infectious disease transmitted by the *Anopheles* mosquito and caused by single-celled microorganisms of the plasmodium genus. The most frequent symptoms are fever, tiredness, vomiting, and headaches. If left untreated, it can be fatal. Several species of plasmodium have been identified, with the deadliest being *Plasmodium falciparum*. The prehistoric origin of plasmodium is in Africa, where it was transmitted to humans by mosquitos from large primates, most likely gorillas.

Malaria was known to all the ancient peoples. The Chinese made reference to it 5,000 years ago and it is described in the *Ebers Papyrus* of ancient Egypt, as well as in Sumerian and Babylonian writings. It found its way to India 3,000 years ago and to the Mediterranean coast roughly 2,000–2,500 years ago. By the beginning of the Common Era, malaria had settled into southern Europe, Arabia, southern and southeastern Asia, China, Korea, and Japan, and was slowly moving north. It had made its way to Spain and Russia by the 12th century, to England in the 14th century, and to central Europe in the 15th century. European conquerors and African slaves introduced it to the New World in the late 15th century. Malaria has been known by various names throughout history, such as ague, intermittent fever,

V. Marko, *From Aspirin to Viagra*, Springer Praxis Books,
https://doi.org/10.1007/978-3-030-44286-6_2

marsh fever, or just plain fever. The Romans divided it into three types according to fever periodicity. If the fever occurred every third day, it was called "tertian"; when it was every fourth day it was called "quartan"; and when the fever occurred every day it was called "quotidian." Malaria is mentioned in Homer's *Iliad*, Shakespeare mentions it in eight of his plays and Hippocrates and Galen both treated it.

The list of famous people who have died of malaria is longer than the closing credits of a Hollywood movie; the names include rulers, popes, politicians, poets, and painters. But mainly, it has killed huge swaths of this planet's population. Some authors have written that malaria is responsible for the deaths of half of all the occupants of this planet.

Story 2.1: The Countess of Chinchón and the Jesuit Bark

Chinchón is a sleepy little Castilian town about 45 minutes southeast of Madrid. Just north of Plaza Mayor, in the center of town on the Plaza del Palacio, stands a small monument. On a white stone pedestal is the bust of a young woman in a lavishly adorned dress and a mantilla over her head. Inscribed below on a bronze plaque are the words:

> EL PUEBLO DE CHINCHON A Dª FRANCISCA ENRIQUEZ DE RIVERA, CONDESA DE CHINCHON, VIRREINA DEL PERU, DESCUBRIDORA DE LA QUINA, EN 1629.

This translates to:

> *The people of Chinchón to Francisca Enriquez de Rivera, Countess of Chinchón, Vicereine of Peru, Discoverer of Quina Bark in 1629.*

Francisca Enriquez de Rivera was the second wife of Luis Jerónimo Fernández de Cabrera Bobadilla Cerda y Mendoza, the fourth Count of Chinchón. Don Luis was from a Spanish noble family close to the Spanish king. As a high-ranking aristocrat, he was named viceroy of Peru by King Philip IV in 1629. The count's wife, Francisca Enriquez de Rivera, accompanied him to Peru and became part of the story of how quinine was brought to Europe. According to legend, she was stricken with intermittent fever in 1638. The governor of Loxa, a province in the south of Peru, learned about her illness and personally brought her a miracle substance that had been used to cure him only recently. He explained to the countess that it was a red powder ground from the bark of a tree that grew high in the mountains where he lived. Francisca recovered and was so grateful that she gave the miraculous powder to everyone who was ill. When she returned to her native Spain in 1639, she brought supplies of the red powder with her and that is how the powdered bark of the *quina* tree was first brought to Europe, as the "countess's powder."

With this noble act, the countess found her place not only in history (for which the grateful residents of Chinchón built a monument), but in botany too. Carl Linnaeus named the tree from which the bark is obtained *Cinchona* in her honor. He left off the first *h* from the name because it sounded better to him without it.

This neat little legend about quinine circulated for many years, but it was in fact just a story. The diary of Don Luis, the count and viceroy, was discovered in the early 20th century and it indicates that his wife was never stricken with malaria and instead had been in good health their entire time in Peru. Moreover, she could not have brought the bark to Europe as she never returned home, dying in the port of Cartagena, Colombia, during the voyage.

There is another legend – more specifically, there are three versions of another legend – associated with the discovery of the benefits of cinchona bark. The legend goes that a fatally ill (a) Indian; (b) Jesuit missionary; or (c) Spanish soldier, drank from a shallow pool of water. He fell asleep and when he woke up, he was cured. All he remembered was that the water he drank had been very bitter – there were fallen cinchona trees lying in it.

So, where is the truth?

The truth is that Jesuit missionaries brought cinchona bark back to Europe with them from Peru in the late 16th century. They had noticed that the Incas, the original inhabitants, drank a broth made from cinchona bark when they had shivers from the cold. They made the connection that the same ingredient could be of use for the shivers caused by malaria, and they were right. The first written mentions of the effectiveness of cinchona bark in treating malaria appeared in 1630. The first samples of the miraculous powder began to appear in Europe around the same time.

It first appeared in Rome, the eternal city surrounded by marshland and bogs, where malaria was so pervasive that they desperately needed a treatment. Malaria was also known as Roman Fever; it was a part of their daily lives, impacting the history of Rome in various eras. The word ‘malaria’ originates from the Latin phrase *mala aria*, meaning bad air. Considering that Rome was the center of political, religious, and cultural life in Europe, Roman Fever also had an impact on European history. The Roman Fever epidemics in the fifth century AD contributed to the fall of the Roman Empire. A half-century later, malaria was one reason why Pope Gregory IV moved the celebration of Hallowmas (All Saints Day), from May when it had normally been celebrated, to November as we still celebrate it today. It was colder in November and there was less risk of a malaria outbreak. There were fewer cases of malaria in the pilgrims who flocked to Rome to celebrate Hallowmas and this reduced the chance of it spreading when they returned home. By the 17th century, malaria had become endemic in Rome. During the papal conclave in 1623, six cardinals died of malaria. The new pope himself, Urban VIII, contracted malaria and it took all his strength to be able to carry out his papal duties. Thirty years later, during the papal conclave that elected Pope Alexander VII, not one cardinal died, because the miraculous powder had come to Rome.

These good results prompted the Jesuits to start importing large quantities of cinchona bark from Peru and distributing it all around Europe. The circumstances were favorable. In the late 1640s and early 1650s, three general councils of the Jesuit congregation were held. Representatives from the entire Old Continent attended and each of the brothers took home a small amount of the bark. Little by little, cinchona bark – as Jesuit's Powder, Jesuit's Bark, and later as Peruvian Bark – was distributed throughout Europe.

If anyone should have a monument anywhere in honor of spreading this new drug across Europe, it would be Cardinal Juan de Lugo and it would be in Madrid.

Juan de Lugo y Quiroga was born in Madrid in 1583. He was a precocious child, who could read by the age of three and was taking part in public religious discussions at the age of 14. He joined the Jesuit order and studied theology at the University of Salamanca. He was a professor at Valladolid and was one of the most important European theologians of his time. Juan de Lugo became part of the history of quinine in 1643, around the time he became a cardinal. He learned about the miracle cure made from cinchona bark from his Jesuit brothers who had brought it over from Peru, and as he had suffered bouts of malaria himself for some time, he became a fierce promoter and zealous defender of Jesuit's Powder once he discovered how much it helped him. He recommended it to the physician of Pope Innocent X, and is responsible for the addition of cinchona bark powder to the *Schedula Romana*, the official formulary of Rome, in 1649. The *Schedula Romana* contains precise instructions for administering the "*bark called fever bark*":

> *"...2 drachms (1/4 oz or 8 g) [of the finely ground power] in a glass of strong white wine three hours before the fever or as soon as the first symptoms occur."*

Juan de Lugo himself bought a large quantity of the bark and distributed it to everyone who was in need of it. In Rome, Jesuit's Powder became known as Lugo's powder (*Pulvis Lugonis*). Because of Juan de Lugo, 1655 was the first year in which not one citizen of this holy city died as a result of malaria.

Despite its indisputable success, Jesuit's Powder did not always receive a positive response in Europe. The principle of its action was unknown, the dosage and treatment duration were only estimated, and the treatment was not always successful. Take Archduke Leopold Wilhelm of Austria, for example. When his malaria symptoms returned despite taking the treatment, he strictly refused to take any more of it. Rather than increasing the dosage, his physician discontinued the treatment because it was deemed useless. The patient died of complications of malaria in 1662.

Nonetheless, cinchona bark as an antimalarial gradually took hold in Europe. In 1650, the Italian physician Sebastian Bado declared that "*This bark has proved more precious to mankind than all the gold and silver the Spaniards obtained from South America.*"

Story 2.2: The successful charlatan and the miracle medicine

We saw at the end of the previous story that cinchona bark gradually found its place as an antimalarial drug across Europe. Now we need to be a bit more precise: it found its place only in Catholic Europe. England, especially in the low-lying areas, was in the grip of malaria to the same extent as Rome was with its marshes, and just like in Rome before Jesuit's Powder was discovered, the main treatment methods for malaria were purgatives, bloodletting, and the occasional trepanation – drilling a hole into the skull. When the first reports appeared about the guaranteed antimalarial drug, this time called Peruvian Bark, most of the Protestants rejected it. They called it the "Jesuit plot" or the "popish powder." One of those who refused treatment was Oliver Cromwell and his death in September 1658 is attributed to untreated malaria, even though this occurred ten years after cinchona bark was officially recognized in Rome as an antimalarial treatment and entered into the *Schedula Romana*.

Fast-forward ten years. In 1668, Robert Talbor, an apothecary apprentice in Essex, developed his reliable cure for malaria.

Robert Talbor was a quack, but a clever one, and he was luckier than most. He was born in 1642 to a respected and educated family – his father was a bishop and his grandfather was the registrar of the University of Cambridge. The young Robert had all the prerequisites to follow in their footsteps, except that he opted for a different life. He dropped out of school in Cambridge and became an assistant at university for a while before working in a pharmacy. As a man small in stature but big in ambition, he decided to study nothing less than the cause and cure of malaria. Looking for a place where "*the agues are the epidemical diseases*," he went to Essex in 1668, where he developed his secret formula. He refused to disclose the recipe, saying only that it consisted of two domestic and two foreign ingredients. This is where the incredible journey of his life began. At the beginning he was a poor, ambitious young man; by the end he was a successful and rich one, physician to two kings, English knight, celebrated healer of the royal courts in London, Paris, and Madrid, and owner of a generous pension. He achieved all this in 13 short years, dying in October 1681 at the age of 39.

His lucky day came early in June 1672 in Essex, where he was practicing. Nearby, there were some sailors recuperating, having defeated the Dutch in Sole Bay just a few days earlier. One of them was a young French officer who was obviously experiencing attacks of malaria. Funnily enough, England and France were allies at the time and the young Frenchman was due to report for duty at the Court of King Charles II, but his malaria made that impossible. The name of this officer has been lost, but the story he wrote in his own hand was preserved. He wrote about how an inconspicuous poor man had come to visit him, bringing with him a powder mixed in with a glass of wine which he told him to drink three times over 24 hours. The quick cure worked so well that the malaria subsided and the young officer was able to report for duty. He was so excited by his miraculous cure that he related the entire event to the king. Charles immediately had the remedy verified, and he was equally as excited about it as the officer. He even appointed Talbor as his royal physician and invited him to come to London. The Royal College of Physicians took offense to the promotion of such an unqualified person, but for Talbor it was the beginning of a great career.

In London, Robert Talbor wrote a slim publication titled *Pyretologia: A Rational Account of the Cause and Cure of Agues*. He specifically warned against the unskilled use of the so-called Jesuit's Powder, observing, "*...as it is given by unskillful hands...most dangerous effects follow...*" He never actually condemned the drug: "*...for it is a noble and safe medicine, if rightly prepared and corrected, and administered by a skillful hand...*" This is an opinion that could be backed even today, had it not been followed by a treatise in which Talbor claimed himself to be the expert with the right hands.

Robert Talbor's luck continued to hold. In 1678, Charles II knighted him and shortly thereafter dispatched him to Madrid to help his niece, Maria Luisa, who was to marry King Charles II of Spain. The wedding was in jeopardy because Maria Luisa suffered an attack of malaria and the strong emetic administered by the Carmelite nuns was not having the desired effect. But the wonder drug of Robert Talbor – now Sir Robert Talbor – was effective and the wedding, which would secure relations between the two most powerful nations at the time, France and Spain, could proceed as planned.

Around that time, Charles II (the English one) also fell ill with malaria. Robert Talbor was already on his way to Madrid, but the king could not wait for his return and self-administered the treatment. It cured him and solidified his belief in the effectiveness of his protégé's remedy.

Maria Luisa, now queen, brought a huge entourage of ladies-in-waiting and servants with her from Paris to Madrid. This was not customary in the Spanish royal court, and a large part of this entourage was repatriated to France soon afterwards, including Robert Talbor. He returned just at the right time. In Paris, the Dauphin, the eldest son of Louis XIV and heir to the throne, had become ill with malaria. Talbor and his miracle tincture came to the rescue.

Louis XIV was thrilled with the miracle cure and immediately wanted to know the secret ingredient. He offered to buy it for 2,000 gold livre (French pounds) and a substantial lifelong pension, an offer not to be refused. However, Robert Talbor had one condition: that the ingredients of the remedy would only be made public after his death. The king agreed – and he did not have long to wait. Sir Robert Talbor died two years later, in 1681. He did not enjoy his royal pension for very long.

Robert Talbor was a quack, but without him, who can say how the history of three European thrones – England, France, and Spain – would have unfolded. And without his miracle remedy, who knows how malaria treatment would have developed in Europe. He was very lucky not to have been a medical authority because his approach to treatment took no account at all of the theories of Hippocrates or Galen. He was, however, a person who discovered what worked – maybe purely by accident – and rigorously put it to use. The difference between him and official authorities may best be exemplified by his response to the question of the cause of malarial fever. His answer was:

"Gentlemen, I do not pretend to know anything about fever except that it is a disease which all you others do not know how to cure, but which I cure without fail."

And what was the secret ingredient of Robert Talbor's miracle remedy? Jesuit's Powder – the ground bark of the cinchona tree.

Story 2.3: The two friends and the yellow cinchona

Pelletier and Caventou. Each of these two men had his own life and his own discoveries; each of them was an important scientist in the first half of the 19th century. But history knows them as "twins" even though the work they did together covers only a small part of their lives, less than five years.

We will begin with the eldest. Pierre-Joseph Pelletier was born in Paris in 1788. His father, Bertrand Pelletier, was a noted 18th-century French chemist and the young Pierre-Joseph followed in his footsteps. He completed his pharmacy studies at the age of 22, was an assistant at the *École de Pharmacie* at 27, and became a professor at the same school at 37. More important for this story is that in addition to his academic life he was also involved in practical pharmaceutics and owned a pharmacy in Paris. The *Pharmacie Pelletier* at 48 Rue Jacob in Paris exists to this day and this is where most of his scientific work was done.

The early 19th century had much to offer in the way of unexplored substances for a perceptive and hard-working young man. Hundreds of new plants had come to Europe in the previous centuries from the Far East and the New World, each one containing as yet unknown substances. Also, there were many plants, both overseas and in Europe, that had long been known and domesticated but still offered new things ripe for discovery. In the back room of his pharmacy, the young Pierre Joseph analyzed and isolated plant resins, plant coloring matter, active substances, and alkaloids.

Pelletier began publishing the results of his analyses in 1811, and his papers came out quickly, one after another. Over the course of six years, he described the composition of resins from more than ten plant sources and he isolated the coloring matter from sandalwood and turmeric. His early work culminated in 1817 with the isolation of emetine, an alkaloid that is still used in pharmaceuticals and biochemistry. In the 1820s, he continued with the isolation and analysis of substances from opium, including morphine. He isolated caffeine from cocoa beans and extracted the active ingredient from curare. The compounds he isolated, analyzed, and described number in the hundreds and he did all this in a relatively short amount of time, dying of colon cancer at the age of 54 in 1842. Pierre-Joseph Pelletier was a highly regarded scientist and a member of renowned French scientific institutions – the Royal Academy of Medicine (*Académie Royale de Médecine*) and the Academy of Sciences (*Académie des Sciences*).

The other "twin," Joseph Bienaimé Caventou, was born in 1795, seven years after his older colleague. He, too, followed the example of his father, a notable pharmacist at the time. He briefly worked as a military pharmacist, but after the defeat of Napoleon at Waterloo he returned to Paris. At the age of 21, before he began to collaborate with Pierre-Joseph, Caventou published a chemical terminology handbook. He also did chemical analyses of plants. After his collaboration with Pelletier ended, he moved on to the chemical testing of several illnesses, including tuberculosis. He was a professor of toxicology and one of the leading experts in his field in France. Like his older colleague, he was also a member of the *Académie des Sciences*. He outlived Pierre-Joseph Pelletier by 37 years.

The number of their individual successes, or their achievements in collaboration with others, would themselves be enough for these two scientists to have a dominant place in the history of chemistry, pharmacology, and toxicology. Their results may have been significant for their time, but they were not enough to put these scientists on postage stamps, have a monument built in Paris, or for one of them to have a crater on the moon named after him. No, that would require their brief collaboration – and mainly the isolation of one single substance.

Their collaboration took place over a relatively brief period of time, from 1817 to 1821, during which they made their essential discoveries in the back room of the pharmacy on Rue Jacob. In that short time, they isolated, named, and described several substances that we still know today.

The first was a green pigment found in most plants. They isolated and described it, and named it chlorophyll, the name derived from the Greek words *chloros* ("green") and *phyllon* ("leaf"). They then isolated the alkaloids strychnine, brucine, and veratrine. Being good researchers, they promptly tested the isolated substances on rats. All three proved to be highly toxic which, apart from other uses, found them a home in mystery novels.

In 1820, they began to study extracts from the bark of several varieties of the cinchona tree. They went on to describe the active principle from six varieties, among them the grey cinchona (*Cinchona condaminea*), yellow cinchona (*Cinchona cordifolia*), and red cinchona (*Cinchona oblongifolia*). The substance that brought them fame was found in the bark of the yellow cinchona (quina).

They named it quinine, and it became the first drug used to treat one of the most serious illnesses, malaria. The isolation process was not easy, but it is worth relaying in this story because it shows the level of knowledge and laboratory techniques available in the first half of the 19th century.

As we already know, the source of quinine was a genus of the cinchona tree known as the yellow quina. The bark was ground and the resulting powder was mixed with alcohol. The alcohol extract was rinsed with a hot solution of hydrochloric acid. This resulted in a purified extract that was mixed with magnesium oxide. The resulting precipitate was then washed and dried, yielding a white powder. The next-to-last step was to dissolve the powder in alcohol. The undissolved part was separated and the clear alcohol extract was evaporated. What remained was a yellow, oily substance – quinine. The name was based on the native name of the cinchona tree, *quinquina* (sometimes it can be found as *quina-quina*). Interestingly, if the grey species of the cinchona tree is used instead of the yellow, the result is not quinine but another alkaloid, cinchonine. Both alkaloids, quinine and cinchonine, are present in the red cinchona tree.

Pelletier and Caventou refused to patent their discovery and instead invited their colleagues to verify the therapeutic properties of the new medicine as soon as possible. They gave up a fortune in doing so. In the 19th century, malaria threatened one-half of the world's population and holding a patent on the process of preparing quinine would have made both of them very rich men.

Before long, the method of isolation, which they simplified several times, became well-known and was used by many pharmacists. In a letter to the members of the Royal Academy of Sciences in 1827, the authors wrote:

> *"...in France...will exceed 90,000 ounces of sulphate of quinine, and admitting that the mean quantity administered to each of those who have taken it, is 36 grains in various doses, ...in 1826, the number of 1,444,000 individuals who have partaken of this remedy."*

For their discovery, the academy awarded them 10,000 francs. It was the only money they would ever receive for the discovery of quinine.

But history more than made it up to them. In 1900, a monument was erected in their honor on boulevard Saint-Michel. After it was destroyed during the Nazi occupation, it was rebuilt in 1951. It depicts a reclining woman atop an eight-foot stone pedestal. In 1970, on the 150th anniversary of their discovery, two postage stamps were issued bearing their images, one in France and the other in Rwanda, while Joseph Bienaimé Caventou has had a crater on the moon named after him. It is 2.8 km in diameter and 0.4 km deep, located in the western part of the Mare Imbrium.

Even in the era of modern synthetic procedures, quinine remains one of the few medicines to be manufactured by the classic method of extraction from the bark of the cinchona tree. Today's method is not that much different from the procedure described in 1820 by Pelletier and Caventou. That does not mean there are no synthetic procedures available; they are just more expensive than the bark extraction method.

There was another attempt at producing synthetic quinine that deserves mention even though it was unsuccessful, because it still went down in the history of chemistry. It was the work of a young man we have already had the pleasure of meeting earlier in this book – the teenage William Henry Perkin. He was the first to prepare the synthetic dye mauveine, which started off a new branch of industry in 1856. The young Perkin had not originally planned to create a purple dye; he had been trying to prepare a substantially more important compound, quinine. The reaction was not exactly how he had imagined, and instead of the expected white quinine crystals, the substance that remained at the bottom of his beaker was a black solid. He wanted to toss it away, but fortunately he first dissolved it in alcohol. A beautiful purple solution was created and mauveine was born. It was probably one of the most successful mistakes in history.

Story 2.4: The unlucky adventurer and the alpacas

It had been a simple arrangement. On one side of the table were the clients: Sir Charles Augustus FitzRoy, governor of New South Wales; the Colonial Secretary, Sir Edward Deas Thomson; and Thomas Sutcliffe Mort, a notable Australian merchant interested in the wool trade. On the other side: one Charles Ledger, an adventurer from England who was living in Chulluncayani, Peru, as an alpaca farmer. Alpacas are similar to llamas, living on the slopes of the Peruvian Andes at elevations of 11,500–16,500 feet. They are smaller than llamas and are raised mainly for wool.

The men around the table were discussing bringing a herd of alpacas from Peru to Australia to raise and breed in a new homeland. The clients promised the supplier a reasonable fee, including around 9,800 acres of land to set up a farm. The exact date is not important, but it was somewhere around the year 1853.

Even as he signed the contract, Charles Ledger knew that while the task may have seemed simple, it would not be at all easy to carry out. The export of alpacas from Peru to Europe was strictly prohibited and punishable by up to ten years in prison, but he relied on his 20 years of experience living in South America. Ledger had come to Peru from his native England as a young man of 18. He eventually became a trader in wool, skins, bark, and copper, and began breeding alpacas in 1848. He was also tenacious and resourceful, and he knew how to bypass the prohibition.

Ledger bought the animals, hired 11 South American herdsmen, and embarked on a 6,800-mile journey across the Andes. He went from Peru to neighboring

Bolivia, and from there to Chile, where he continued south to reach the Pacific Ocean at Copiapo. The journey took nearly six years. He was forced to replenish the herd along the way, care for them in the winter, search for suitable pastures in the summer, and protect them from the local natives, but his journey was successful. They boarded ship in Copiapo, and on November 28, 1858 he disembarked with 256 animals and the 11 South American herdsmen on the Australian coast.

The six years it took Charles Ledger to travel across South America was a pretty long time, and in the meantime a lot of things had changed in Australia, especially the clients. The governor, Sir Charles A. FitzRoy, had ended his tenure and returned to England; the Colonial Secretary, Sir Edward D. Thomson, had taken up another position; and Thomas S. Mort, the merchant, was recovering from the huge losses he had suffered in financing a massive shipyard. Suddenly, nobody was interested in breeding alpacas in Australia. The New South Wales government paid Ledger £15,000, but that was only enough to cover his costs. He received no land and no one was interested in his alpacas. He tried to sell them at auction in 1863, but to no avail. He was left with no other option but to give up. He wrote, "*On the faith of promises made in this country, I undertook every risk – did succeed – and am ruined!*" It is unclear what happened to the animals, but at the beginning of the 20th century there was not one alpaca in Australia.

This failure might have broken another person, but Charles Ledger was an adventurer and in 1864 he was not ready to sit around and do nothing. He decided to try his luck again, this time choosing a completely different commodity. He decided to export cinchona seeds from South America.

After the two young friends, Pelletier and Caventou, had discovered the process for extracting quinine from the bark of the cinchona tree, the demand for quinine skyrocketed. Bark traders in South America were willing to pay good money for it, so the native suppliers – *cascarilleros* – felled one tree after another. Long gone were the times in the 17th century, when the Jesuits had taught their great-grandfathers to plant five new trees for every one they chopped down. They had taught them to plant the new trees in the shape of a cross because they believed that this would make the tree god help them grow faster.

Cinchona trees began to disappear and it became more difficult to find species with sufficient quinine content in the bark. Moreover, the South American countries stringently guarded the bark trade and would not let foreign traders participate. The two main colonial powers, Holland and England, were not at all happy with what was happening. They wanted to move the cultivation of cinchona trees to their colonial territories. The Dutch wanted to cultivate the tree in Java, Indonesia, while the English wanted to cultivate it in India, and thus began a race to see who would be the first to establish cinchona plantations outside South America.

It was not at all easy. Cinchona trees grow primarily in inaccessible tropical forests on the western slopes of the Andes, from Colombia in the north to Bolivia in the south, at elevations of 5,000–10,000 feet. Furthermore, not all of the several dozen species have enough quinine in their bark to make them usable for production. The first attempts to export seeds and saplings from Peru and relocate them to new areas were not successful. Some attempts failed while still in South America; in others, the saplings did not survive the long journey. When there was finally a successful journey, the seeds often did not sprout in their new homes. Even where the saplings managed to grow, the quinine content in their bark was not high enough. The Dutch and English attempts to domesticate cinchona trees ended in failure.

Enter Charles Ledger. Apart from having tenacity and persistence, he also had Manuel Incra Mamani. Manuel was a Bolivian *cascarillero* and Charles's loyal servant and companion. He was an expert like no other in the cinchona tree and knew the various species by the shape and size of the leaves, and by the color of the flowers and bark. He knew the places where they grew and the places where it was pointless to look. Ledger charged him with combing the Bolivian rainforests for suitable trees and collecting as many seeds as he possibly could. It took him five years. When he returned to civilization, he brought two sacks of cinchona seeds for his master, weighing a total of about 45 pounds. Cinchona seeds are very small; one seed weighs less than 3 mg. The faithful Manuel Mamani had gathered over six million seeds.

Ledger carefully dried the seeds, hid them among some chinchilla pelts, and sent them to his brother in London. The rest of the tale is rather tragicomic.

As soon as Charles's brother received the package, he immediately took it to the Royal Botanic Garden, assuming he would be met with great interest. They rejected his offer. He went around to several other potential customers, from the manager of another botanic garden, to an important member of a British pharmaceutical society, to a cinchona trader. Everywhere he was met with rejection, and he began to panic. The seeds could lose their germinability and the efforts of Charles Ledger and Manuel Mamani could all have been for naught. But fortune smiled on him and he found a buyer willing to purchase about a pound of seeds. It was the Dutch consulate general in London, and the price was 100 Dutch guldens (guilders), about £50 at the time. That was it. That was all that was used out of the entire consignment of seeds. The rest either remained unsold or never sprouted in their new home. However, the majority of the seeds bought by the Dutch took root in Java and it turned out that the species gathered by Manuel for Charles Ledger had a much higher quinine content than any species before it. An English adventurer, his Bolivian servant, and the disinterest of the English, gave the Dutch a monopoly over quinine. For many years, Holland ruled the quinine trade and earned unprecedented profits. In 1900, two-thirds of the world's quinine came from the Java plantations. Since then, the purchase of cinchona seeds for 100 guldens has been considered one of the best trades in history.

The adventurer Charles Ledger died in Australia in 1905. He died in poverty but had lived to the respectable age of 87. His servant, the *cascarillero* Manuel Incra Manami, was far less fortunate. Shortly after he delivered the sacks of cinchona seeds to his master, he was charged with their illegal export. He was sentenced to a beating, and because he refused to say who he had been working for he was beaten over and over. He died a few days later as a result of his punishment.

Charles Ledger and his impact on the propagation of cinchona trees did not remain forgotten. In 1994, a memorial was erected in his honor in Sydney, and the species *Cinchona ledgeriana* was named after him, as was a type of quinine beverage – *Ledger's Tonic*.

But there is no memorial in honor of his faithful servant, Manuel. It was Manuel who spent five hard years gathering seeds in the rainforests of Bolivia. It was Manuel who paid for it with his life.

History is not always just.

Story 2.5: The two opposing scientists and the mosquitoes with spotted wings

This day relenting God
Hath placed within my hand
A wondrous thing; and God
Be praised. At His command,
Seeking His secret deeds
With tears and toiling breath,
I find thy cunning seeds,
O million-murdering Death.
I know this little thing
A myriad men will save.
O Death, where is thy sting?
Thy victory, O Grave?

This poem was written by Ronald Ross on August 20, 1897, the day after he discovered a malarial parasite inside the stomach of a mosquito.

For a change of pace, this story is not about a drug but about the illness a drug treats, malaria. The story of how malaria parasites were discovered is so colorful and full of adventure it deserves to be told. The history is long and there are several associated names. Two men claimed to have discovered the malarial parasite – the Englishman Sir Ronald Ross, and the Italian Giovanni Battista Grassi. These two could not have been more different.

Ronald Ross, later Sir Ronald Ross, was so controversial that you would be hard pressed to find another person like him. He did not want to be a doctor, and when he became one, he was less than enthusiastic about performing a doctor's duties. He nonetheless earned a place in history as an indefatigable medical experimenter. Ross was a true researcher, willing to push past his comfort zone in the pursuit of his goal. He was ambitious, egocentric and contentious. He was also self-conscious and insecure. He could not have conducted his research without the professional and emotional support of his teacher and mentor.

If anyone ever deserved to be called a chronic complainer, it was Ross. He complained about everything. He criticized governmental agencies for lack of funding. He complained to his employer about low pay. He was also unhappy with his lot in life – and yet he was knighted, was awarded the Nobel Prize, and was a member of select European science societies. A science institute, a university building, and streets in numerous cities are named after him.

Ross was born in 1857 in India, to a father who was an English general serving in the colonial army. The father must have been a striking individual; even his son, Ronald, as an adult looked like a colonial general on vacation. As a proper young gentleman, Ronald was sent to England to attend school and he did well at first. At the age of 14 he took first place in a math contest and two years later he won a drawing contest. He could draw a copy of a painting done by the Renaissance master Raphael in just a few minutes. He wanted to be a painter, poet, composer, and mathematician, but was unsure whether it would be better to excel in these areas one at a time, or in all of them at the same time.

However, his father, the general, had a different future planned for his son. He wanted his son to be a military doctor, and so Ronald began his studies at St. Bartholomew's Hospital Medical College in 1875. He was not an exemplary student, but he squeaked through and graduated in 1879. Two years later, he took the special exam that allowed him to enter the Indian Medical Service, fulfilling his father's dream of having a military doctor for a son. He served in India, and for the next seven years nothing extraordinary occurred in his life. By 1888, he had become bored with the routine work and was very happy to take a one-year leave in London. During his leave, Ross wrote another failed novel, developed a new system of shorthand, invented phonetic spelling for verse writing, became the secretary of a golf club, earned a diploma in public health, and learned the essentials of microscopy. He also married the dark-haired Rosa Bessie Bloxam.

In 1899, he returned to India and continued the routine work of a military doctor. As a diversion, he decided to formulate a theory about malaria. Not that a similar theory did not already exist; in fact, there were several.

The second half of the 19th century was a time of advances in bacteriology, and malaria was an illness that had been classified as one of those caused by bacteria. It was thought that the illness was caused by bacteria found near stagnant, dirty water and entered the human body by inhalation. That was how various other bacterial illnesses were spread, and there was no reason to think malaria was any different.

The French epidemiologist, Charles-Louis-Alphonse Laveran, opposed that theory. He had discovered foreign bodies in the blood of patients infected with malaria that were unlike bacteria and seemed a bit strange. They had various shapes and underwent transformations, multiplied, and often disappeared. The discovery was so unusual for the time that the official scientific community unanimously rejected it.

Of course, Ronald Ross would not have been Ronald Ross if he had not formulated his own hypothesis of malaria right from the start. He had a microscope he had learned to use in London and enthusiastically set out to refute Laveran's findings. He drew the blood of malaria patients and spent hours looking for Laveran's strange bodies. He found nothing. The only possible conclusion was that Laveran was wrong and no microscopic bodies existed. Ross believed malaria was caused by bowel poisoning.

In 1894, after five further years of service, he was allowed another leave. That year would be a turning point in the life of Ronald Ross. On the recommendation of the colleagues with whom he had shared his hypothesis that no microscopic bodies existed, he visited a renowned expert on tropical diseases, the Scottish doctor Patrick Manson, who was working in London. This turned out to be a good decision. Dr. Manson was not only an expert on tropical diseases, he was also a mosquito expert. In fact, he was obsessed with the insects. He believed that mosquitos played a much greater role in human life than was thought. Everyone had their doubts, but it turns out he was right.

Under Manson's microscope, Ross saw for the first time the crescents described by Laveran in the blood of a malaria patient. They were later given the name plasmodia. The two men spent many hours together in Manson's laboratory and taking walks. They were quite a mismatched pair. Patrick Manson would come to be known as the father of tropical medicine. He was a gentle man with a typical Scottish sense of humor. Ronald Ross was completely different. He was an ambitious, over-confident, egocentric military doctor who at first did not even know that mosquitos are a type of fly. But Manson had faith that with the help of his younger colleague, he would be able to prove that mosquitos were responsible for the propagation of malaria.

In March 1895, Ronald Ross was back in India serving in the Indian Medical Service, but this time he was on a mission. He wanted to prove his teacher's hypothesis. Between 1895 and 1899, they exchanged 173 letters, all of which have been preserved to form a unique trail of scientific correspondence that lays out Ronald Ross's journey to the Nobel Prize, step by step.

Ross knew that if he was to prove Manson's theory, he would have to find plasmodia not only in the blood of patients but mainly in the stomachs of mosquitos. Success eluded him for a long time. He caught mosquitos and released them onto malaria patients under mosquito netting; he killed, dissected, and observed mosquitoes under a microscope, all to no avail. His letters to Manson were charged with desperation and hopelessness. Manson was patient, guiding Ross and showing him new ways, but what he mainly did was to encourage his younger colleague during his bouts of depression.

At the time, neither Ross nor Manson knew that there were several hundred species of mosquito in the *Anopheles* genus and that only a small percentage of them could transmit the dangerous plasmodium parasites from human to human. The most dangerous of these is the *Anopheles gambiae.* The mosquito with the spotted wings.

The watershed moment occurred in August 1897, when his assistant brought him just a few mosquitos, not more than 20. They stood out for the fact that their wings were spotted. For a small reward, Ronald Ross convinced one of his malaria patients, Husein Khan, to let the mosquitos bite him. On August 19, 1897, Ronald Ross began dissecting the mosquitos and observing their stomach contents under the microscope. Finally, he saw the small, circular, dark-pigmented microorganisms, similar to those he had already observed so many times in the blood of patients. He was the first person to see malaria-causing plasmodia in the stomach of mosquitos, finally proving his teacher's theory that mosquitos transmit malaria. He wrote his poem the very next day as a tribute to his discovery.

All that remained was to find out how to transmit the plasmodia from the mosquito to a human. Just as he was about to complete his research with success, Ross was transferred to another part of India where malaria was not prevalent. Once again, Manson came to the rescue. If Ross could no longer study malaria in humans, why not continue the research by studying malaria in birds. Mosquitos must be the vector here as well, and it must also be plasmodia that were responsible for the transmission of the disease to birds. Ronald Ross's laboratory was soon transformed into an aviary full of sparrows, larks, crows, and other birds. In July 1889, less than a year after his discovery of plasmodia in the stomach of mosquitos, he made a final important discovery. He found plasmodia in the salivary gland of the mosquito. He thus closed the circle that begins with the blood of a sick human, continues in the stomach of the mosquito that sucked that blood, and ends in the blood of a person who had been healthy until that point through the saliva of the same mosquito.

Giovanni Batista Grassi was the other to lay a claim to the discovery. At the beginning of this story, we learned that he was the exact opposite of his rival. He was a trained doctor and zoologist, professor of comparative zoology at the University of Catania, and later professor of comparative anatomy at the University of Rome. One of his biographers described him with the words "*deliberate as a glacier, precise as a ship's chronometer.*" His approach to determining the cause of malaria was completely different from that of Ronald Ross. Grassi is proof of the accuracy of one of the most fundamental truths in science and research: that half the answer is in asking the right question.

His question was: how can some places be infested with mosquitos and have no malaria, when all the places where there is malaria are infested with mosquitos? There was only one answer to this question. If mosquitos are responsible for malaria, then there must be mosquitos that transmit malaria and other mosquitos that do not. How simple.

So Grassi traveled around Italy, catching mosquitos and separating them into two types. One type was safe, the type caught in places without malaria. The other type was suspicious, having been discovered in places where malaria was endemic. He ultimately ended up with one suspect, an elegant mosquito with dark spots on light brown wings.

The first step was to prove that these mosquitos could transmit malaria from an infested area to an area where malaria was not abundant. Professor Grassi took a jar full of mosquitos caught in the infested areas and brought them to the clean environment of the knolls of Rome. He found a group of volunteers who had never had malaria and who were willing to undergo an itchy experiment. Nearly all of them very soon came down with a fever and shivers. It was a pretty risky experiment; quinine, the cure for malaria, had been around for quite some time, but one never knows...

The time for definitive proof came in 1900, with one simple clinical experiment. Giovanni Grassi selected a settlement of railway workers in an area where malaria was most abundant and divided them into two groups. One group, consisting of 120 workers and their families, had all the doors and windows of their homes covered with fine mesh screens. The residents of these homes were not allowed outside after sunset, when the mosquitos come out to hunt. The second group, consisting of 415 people, lived as they normally did without protective screens on their homes and were allowed to go out after sundown. The results were clear: nearly all the people in the second group contracted malaria. Only five in the first group got sick.

What Ronald Ross had demonstrated with bird malaria, Giovanni Battista Grassi confirmed in humans. Furthermore, he was an experienced zoologist, providing a precise description of the culprit, including its gender. He was the one to discover that only the females were dangerous.

Both Ross and Grassi were subsequently nominated for the Nobel Prize in physiology in 1902, but at the time no one was aware of one of the most bitter scientific feuds in history. It started with Ronald Ross, in a manner that was all his own. He referred to his rival as a charlatan and a cheap crook, a parasite who survived on the ideas of others. Grassi responded in a similar tone, and the "scientific discussion" continued along the same lines. It seemed as though these two notable researchers would sooner give up their award than let their rival have it.

The truth of the matter is that the zoology professor, Giovanni Battista Grassi, had somehow forgotten to give credit in his paper to an ordinary military doctor from India. Yet many of his findings were based on Ross's work. It is also indisputable that Grassi's contribution was really just a conclusion to Ronald Ross's findings. On the other hand, Ronald Ross would probably never have reached his conclusion without the help of his teacher, Patrick Manson. Not forgetting that the work of both was based on the discovery of plasmodium by Charles-Louis-Alphons Laveran.

Only Sir Ronald Ross was awarded the Nobel Prize. In addition, he received many other awards and honors, including his knighthood. Nevertheless, for the rest of his life he felt undervalued and complained about being underpaid for a scientist of his stature. He died in 1932 at the age of 75.

Concluding remarks

Quinine made an enormous contribution to the history of the world. Without it, the Panama Canal would never have been built, tea plantations in India would not have been planted, and offshore oil platforms in Venezuela would never have been constructed. There would be no railroad through the Amazon Forest. Quinine – and its predecessor cinchona bark – played a critical role in the colonization of Africa. Quinine powder or tablets were packed into the bags of all the soldiers, government officials, merchants, missionaries, and everyone else who took part in any way in the settling of the African continent. Thanks to quinine, Africa stopped being known as the "white man's grave."

Today, new antimalarials have taken the place of quinine and it is no longer a first-line drug. It is now much better known as a carbonated soft drink, with a quinine content that is much lower than the amount used in malaria treatment. One eight-ounce can of tonic water contains less than a tenth of the amount of quinine contained in one quinine tablet.

And one last little tidbit. Both quinine and cinchona bark have a very bitter, unpleasant taste. The English who served in India found a way to make it palatable, by mixing it into gin, which is how the gin and tonic came to be a popular drink.

3

Vitamin C

Of all the diseases caused by malnutrition, scurvy has probably caused the most suffering throughout history. In the Age of Sail, it was responsible for more deaths than storms, maritime disasters, military operations and other illnesses combined. Conservative estimates say that scurvy took the lives of more than two million sailors. It is an awful disease, with symptoms that include bleeding gums, tooth loss, fetid breath, old wounds reopening, and once-healed fractures re-breaking. Left untreated, scurvy leads to a slow and inevitable death. While it is usually associated with sea travel, it was regularly seen in the northern hemisphere, especially in the winter months. It accompanied sieges and it was found in prisons; everywhere people lacked vitamin C. The author can still recall the 1950s and 1960s and the annual "spring lethargy." It would show up at the end of winter as moderate symptoms of scurvy. Fresh sources of vitamin C were only available in the growing season at that time and there were fewer canned sources on the shelves in early spring. Only fairytale creatures had strawberries in December.

The eradication of scurvy was a huge advance in medicine, comparable to the discovery of a vaccine for smallpox. The road to this outcome was not in the least bit linear; the cure was discovered, then forgotten, then rediscovered, again and again. There are several names that came up along that road which should not be forgotten. We will meet six of them.

Story 3.1: The famous admiral and scurvy

It was truly a triumphant parade, the likes of which London had not seen in ages. Leading the way was captain George Anson, a sturdy 47-year-old man and future First Lord of the Admiralty. He smiled and waved to the crowd lining the streets

V. Marko, *From Aspirin to Viagra*, Springer Praxis Books,
https://doi.org/10.1007/978-3-030-44286-6_3

of London to honor him, the great hero. Most of his 188 men walked behind the captain. They had all just returned, on June 15, 1744, from a voyage around the world on the *Centurion*. All of them were weather-beaten old salts whose faces did not give away their real ages and they were waving and smiling too, their broad grins revealing toothless mouths; a price paid for a lengthy time at sea.

But Captain Anson and his men were not the only reason the Londoners had gathered in the streets. Bringing up the rear of the parade were 32 wagons full of silver. It is alleged to have been 70 tons of silver, captured by the captain and his men from England's archenemy, the Spanish Crown.

The riches brought home by the *Centurion* completely overshadowed one very serious fact. Captain George Anson's voyage around the world was one of the greatest tragedies in the history of sea navigation in the Age of Sail.

They experienced trouble from the very beginning. It was July 1740, and George Anson had been waiting since February to set sail with his fleet of eight ships. He had to get the wind in his sails as fast as possible in order to carry out the difficult orders given to him by King George II. He was to set out heading west and sail around Cape Horn at the southern tip of South America to reach the Pacific Ocean. From there, he was to continue north along the coast, destroying every Spanish ship he might encounter along the way. But his primary destination was Acapulco, a city on Mexico's Pacific coast, where he was to await a Spanish galleon full of silver on its regular voyage between Mexico and the Philippines. He was to seize the silver in the name of His Royal Majesty. England was at war with Spain at the time and this type of operation was in line with military tactics.

To fulfill his orders, Anson was given five big warships, each with two decks of cannons; one smaller ship with one deck; and two escort supply ships. He had a total of 232 cannons – incredible firepower.

But it was already summer and Anson could not set sail, even though the ships were repaired and ready and the cargo had been loaded, because he did not have a full crew. He needed around 2,000 men and he was still short by several hundred. Service in the Royal Navy was not very popular because it was difficult and grueling, so there was no steady stream of volunteers available. Once aboard a ship, the status of regular sailors was not much different to that of slaves. Discipline was paramount and floggings were the order of the day. The life of a sailor had little value and mortality was high, so ships tended to be overcrowded. Men were stuffed below decks like sardines and their living quarters consisted of a hammock hung just a few inches away from their nearest neighbor. The crew was lice-infested and stricken with all manner of infectious diseases. During rainy periods, there was mold everywhere.

The food was not much to speak of either. The menu consisted of salt pork and fish, flour, oats, cheese, butter, molasses, and dried biscuits. The pork, fish, butter, and cheese began to reek after a while and the flour and biscuits would quickly be crawling with flour moths. That was actually a good thing as far as the biscuits were concerned because it made them more porous and easier to bite into.

The lack of volunteers meant that force had to be used. Groups of four or five burly men armed with heavy sticks were sent ashore to beat and capture men and drag them on board the ship. Even so, they could not find enough crew to set sail, so the Admiralty decided to provide the captain with 500 veterans from the Chelsea Hospital. They were mostly older, sick, and invalided men, and only about half of them were able to stand on their own feet to be recruited; the remainder stayed on shore. They were the lucky ones. Nearly all of the veterans that set sail died before they reached South America. There was another reason the Admiralty wanted to move those men out of the hospital: the beds were needed for the sick and injured sailors returning from the campaigns of the Royal Navy forces.

The final option was to recruit young marines. They were too young and too inexperienced, but they made up the number of men required and the crew was at last complete. Finally, on September 18, 1740, George Anson set sail with his eight ships and 1,854 men aboard, with Anson at the helm of the flagship, the 1,000-ton colossus *Centurion* with its 60 guns.

After briefly dropping anchor in Madeira to replenish stocks, they continued sailing westward. The crew began getting sick and dying on the journey to South America. The *Centurion* alone lost 20 men to typhus, dysentery, and later malaria, while another hundred were incapacitated; but that was just the beginning. After they had been at sea for some time, the most feared disease of sailors began to appear – scurvy.

The food served to the men in the service of the Royal Navy did not contain many sources of vitamin C, which is the name later given to the substance that causes scurvy in humans when there is a deficiency. It also did not help that many of the recruits came on board already suffering from malnutrition. It is little wonder that after several months at sea, this disease struck the crew of each of the ships.

Scurvy had been well-known for a very long time. It is mentioned in the *Ebers Papyrus* from ancient Egypt and in the Old Testament, while Hippocrates also documented the disease. Symptoms begin with malaise and lethargy, then continue with bleeding gums, tooth loss, skin changes, pain and, if left untreated for very long, death.

The savageness of scurvy began to manifest fully at a time when the building of great ships, and the courage of sailors, meant that long voyages could be undertaken without the opportunity to replenish supplies. In 1499, Vasco da Gama lost 116 of 170 crew members, while three years later, Magellan lost 208 of 230. In both cases, the main cause was scurvy. Scurvy proliferated in the Age of Sail. In the 18th century, it killed more Royal Navy sailors than were lost in battle. The English records are precise: during the Seven Years' War from 1756 to 1763, the Royal Navy enlisted

184,899 sailors; 133,708 of them died from disease, mainly scurvy. If we add it all up, around two million people died of scurvy between 1500 and 1800.

But let us return to Captain George Anson's voyage. In March 1741, his fleet reached Cape Horn, the southernmost tip of Latin America, where the very thing that the captain had been worried about as he waited for men and supplies to be replenished began – the season of gales and violent storms in the southern hemisphere. For the crews on deck, this meant a three-month battle for survival. Ships were tossed about like paper boats on the massive seas and just when the crews needed their strength the most, they began to weaken. One-third of the crew was soon incapacitated, and others followed; then they began to die. At first, it was one individual here and there, but then more and more were dying. The *Centurion* lost 43 men in April and twice that number in May and the other ships were no better off.

Two of the ships gave up hope of sailing around the Cape and turned back toward the Atlantic Ocean. A third ship fared even worse. Its crew had been disabled the most by disease and the onslaught of the storms proved too much. Gusts of wind blew it onto the rocky shore, where the ocean literally ripped it to shreds. Most of the crew died, being too weak to save themselves.

Captain Anson was left with only three warships. When the seas finally calmed in May 1741, it was time to count their losses. The *Centurion* barely had 70 men fit for duty, and the other two ships were not in much better shape. The *Gloucester* had lost two-thirds of her crew, and those remaining were barely able to perform their duties. The smallest ship, the single-decked *Tryal*, lost half of her crew. Only four men, including the captain, had the strength to hoist the sails. Of the 1,200 men who had boarded the three ships, only 335 survived.

They spent four months on a small island off the Chilean coast, where there were plenty of fresh vegetables and fish. The sailors had a chance to recover and get back to their task. They attacked six Spanish merchant ships and burned down towns along the coast of Peru, losing only one ship, the *Tryal*. They split the loot, worth over £70,000, among themselves according to the customs at the time.

The Spanish became aware of the activities of the English fleet, and George Anson's war mission found itself in danger. He waited near Acapulco for the silver galleon a while longer, before deciding to sail west toward China in May 1742. Whereas the year before they had faced furious gusts of wind near Cape Horn, in the summer of 1742 on their voyage across the Pacific they faced near wind-less conditions. They made very slow progress and scurvy broke out again. The first death was recorded on July 5 and later the toll began to climb to five dead per day. By mid-August, the situation was critical. The remaining sailors could not handle both ships, and they had to sacrifice the *Gloucester*. The men and cargo were transferred to the *Centurion*, and on August 13, 1742, the pride of the English naval fleet and Anson's second largest ship was set on fire to keep it from falling into the hands of the Spanish.

But the losses continued and more of the crew died every day. Burdened with all the looted silver, the *Centurion* was also in very poor condition and had sprung leaks. The surviving men were so weak that they could not work the pump fast enough so the officers, even Anson, took their turns at pumping. It would only take a few more days for the *Centurion* to sink with all its silver, cannons, and a few hundred crew. Fortunately, they encountered the Mariana Islands with all their tropical vegetation. All those who had survived were rescued, and they even managed to bring the crippled *Centurion* into Macao harbor and patch it up as best they could.

In April 1743, the final act of their voyage could begin – the wait for the Spanish galleon. However, it would not be at the ship's starting point in Acapulco but instead near its destination in the Philippines. It was very risky. The *Centurion* only had 227 men instead of the 400 normally needed to man the ship and the guns. They knew there were not enough men, and they also knew the ship could soon be wrecked. They had lost their teeth, their health and their friends. The prize from such a huge capture would at least in part make up for all their suffering on the voyage, and when the Spanish galleon *Covadonga*, the "prize of all the oceans," was spotted on June 20, 1743, they launched an attack. It only took 90 minutes, and Captain George Anson knew he would return from his voyage around the world a victor.

We know the rest of the story. The voyage brought money and fame to those who survived. Of the 1,854 men who set sail from Portsmouth on September 18, 1740, the lone *Centurion* returned to her home port with only 188 aboard. Together with the others who survived on the ships that had left the convoy back at Cape Horn, the total number of survivors was 500. All the rest perished, the majority from scurvy.

Story 3.2: The ship's doctor and Murphy's Law

> *The discovery of a fact that opposes the theory generates a force that works to reject that fact.*
> Murphy's Law of Research.

Captain George Anson was not the first captain to lose nearly his entire crew to scurvy; unfortunately, he was not the last either. The maritime powers of Great Britain, Spain, and France spent virtually the entire 18th century waging some war or other on various seas and oceans. Captain Anson's expedition was part of a war that went down in naval history with the somewhat ridiculous name of the *War of Jenkins' Ear*. This is the incident that inspired the name. The Spanish coastguard detained British merchant Robert Jenkins and his ship off the coast of Florida on suspicion of smuggling. The captain of the guard, Julio León Fandiño, cut off

Jenkins' ear with the words "*the king would suffer the same punishment if caught doing the same*." Jenkins did not delay, and on returning home to England he informed the king, who took it as a personal offense and proceeded to declare war on Spain. It lasted nine years, from 1739 to 1748.

The War of the Quadruple Alliance came before this war, and before that was the War of the Spanish Succession. After the War of Jenkins' Ear, the Seven Years' War (1756–1763) broke out, and then there was the American Revolutionary War (1775–1783). The warring finally ended, at least for a time, with the Battle of Trafalgar in 1805. The battling armies had hundreds of warships and transport ships with tens of thousands of men aboard. If you add to that the constantly developing trade between Europe, Africa, Asia, and the New World, there were thousands of ships crisscrossing the oceans of the world. And enormous numbers of seafarers were plagued by scurvy; it became one of the first serious occupational illnesses.

It did not have to be that way. Although the cause of scurvy would remain a mystery for a long time, the way to prevent it had long been known.

The first indication that sailors knew about the beneficial effects of citrus fruit on scurvy came from the travels of Vasco da Gama, later confirmed by Pedro Álvares Cabral, in the 15th and early 16th centuries. Portuguese explorers even planted fruit plantations in Saint Helena, an island in the Atlantic Ocean, and sailors ate the fruit to treat scurvy on their voyages from Asia around the Cape of Good Hope. The knowledge of the Portuguese explorers was also used by one of Queen Elizabeth's prominent privateers, James Lancaster VI, who attacked and pillaged ships in the Indian Ocean in the late 16th and early 17th centuries – long before George Anson's voyage. The crew of his ship received a daily dose of lemon juice.

It was not only citrus fruit that saved the lives of travelers. In 1536, on the other side of the world in North America, Jacques Cartier and his men were exploring the Saint Lawrence River area. They probably would have died had the local inhabitants not shared their recipe for a scurvy cure – a potion brewed from an evergreen tree called *Arborvitae*. The needles of this tree contain 50 mg of vitamin C per 100 g, which is about the same amount contained in an orange or a lemon. Information was recorded by other explorers and voyagers of the 16th and 17th centuries, who already knew that scurvy could be prevented by eating fresh fruit.

It was not just sailors and explorers who knew the remedy for scurvy: one was found on the land as well. It was written in 1707 by a Mrs. Ebot Mitchell of Hasfield, Gloucestershire, and is worth noting here (and maybe even trying):

> *A blend of three handfuls of watercress, a like amount of spoonwort and speedwell, one handful of bishop's wort, and half a handful of wormwood is mixed into a liter of wine, covered and left to stand for 12 hours. The extract is poured*

into a bottle. It is mixed with the juice of eight oranges. The afflicted person is to take eight spoons of the mixture daily along with a draught of beer.

An experiment conducted by a Scottish ship's surgeon in the service of the Royal Navy could have ended scurvy and saved many lives. It began on May 20, 1747 aboard the HMS *Salisbury*, a new 50-gun fourth-rate warship with a crew of 350. The experiment was conducted by James Lind, the ship's doctor, and is still considered the to be the first ever controlled clinical trial in medical history.

At the time, the 31-year-old Lind was already a seasoned professional, having served eight years as a surgeon's mate sailing the seas and oceans on various ships. He knew that scurvy was an inextricable part of a sailor's life, so he was not at all surprised when scurvy broke out among the sailors on the *Salisbury* just two months after setting sail from her home port of Portsmouth.

The experiment that immortalized him was fairly simple. He selected 12 sailors afflicted with scurvy and separated them into six groups of two. They all followed the same standard diet, consisting mostly of oatmeal for breakfast, mutton broth for lunch, and a broth of barley and raisins with rice and wine for dinner. Each pair was given a different type of treatment. The first pair was given a quart of apple cider every day; the second was given 25 drops of elixir of vitriol (diluted sulfuric acid) three times a day; the third was given two spoonfuls of wine vinegar three times a day; the fourth was given one-half pint of sea water; and the fifth was given two oranges and a lemon. The sixth pair was given the scurvy treatment recommended in the mid-18th century: a blend of garlic paste, mustard seeds, dried radish, balsam of Peru, and myrrh, to be washed down with barley water. James Lind included the seawater treatment mainly to please the Admiralty because it was a quick and cheap way to treat scurvy. In those days, vitriol was very often used to treat the disease.

Looking at it today, we know which of the six pairs received the effective treatment. James Lind was able to conclude that "*the most sudden and visible good effects were perceived from the use of oranges and lemons.*"

Shortly afterwards, the war between England and Spain ended. James Lind retired from the navy and became a successful private doctor in his native Edinburgh. He continued to study and collect all available information about scurvy and, in 1753, he published his life's work: a 400-page tome with the unconventional title: *Treatise of the Scurvy in Three Parts. Containing An inquiry into the Nature, Causes, and Cure, of that Disease. Together with A Critical and Chronological View of what has been published on the Subject.* The Treatise included a brief recounting of the experiment he conducted aboard the *Salisbury* in May 1747.

By the time the Treatise was published, its author was not unknown. In 1750 he was elected a fellow of the Royal College of Physicians of Edinburgh, and in 1758 he was appointed as Chief Physician to His Majesty's Royal Hospital in Gosport, near Portsmouth on England's south coast. At the time, it was the largest hospital in the country. He dedicated the book to George Anson, First Lord of the Admiralty. Despite all this, the results of Lind's experiment on the *Salisbury* were largely ignored by the medical and naval authorities of the mid-18th century.

There are two schools of thought as to why. The first is a typical example of the view, usually simplified, that it is difficult to advocate a new idea. The roles here are the visionary on the one side and the rigid establishment on the other. This is reinforced by the fact that it was more than 40 years after the Treatise was first published before the Admiralty changed its approach to treating scurvy. It took 40 years for them to recommend lemon juice as the routine cure for the disease on all of its ships.

The second is based on the fact that the mentality of people in the 18th century – and how they approached fact and theory – was completely different from that of today. The ancient theory of Hippocrates formed the foundation of medicine during this time and was the point of reference for the causes and treatment of diseases. It was based on the need for a balance of the four bodily fluids – blood, phlegm, yellow bile, and black bile. According to Hippocrates, the fluids were in balance in a healthy person and all diseases, whether physical or psychological, were caused by an imbalance. The treatment of diseases was based on bringing the bodily fluids into balance. Scurvy was a black bile disease, so it was considered to be dry and cold and could be treated with something that had the opposite properties: wet and warm. Since lemon juice was also considered to be cold, it was not deemed suitable for the treatment of scurvy. It did not matter that the outcome of practical experiments showed something different; the theory could not be contradicted.

The problem was that even James Lind was unable to assess his observations properly. He believed that scurvy was caused by faulty digestion and elimination. In this theory, the gastrointestinal tract breaks down food into small particles. Some are used to renew the body and some are eliminated, especially through perspiration. If food is not perfectly digested and the waste products are not properly eliminated, the body begins to putrefy. On long voyages, sailors endure difficult conditions and are unable to digest food properly. The stomach does not break down the food perfectly and it cannot be eliminated from the body. Additionally, in bad weather the pores close up, further worsening elimination. Bleeding gums, skin ulcers, and fetid breath are definitely signs that the body is putrefying. Diet has nothing to do with scurvy. The theory sounds complicated, but at the time it was quite comprehensible.

Lemon juice did not fit this theory, if only for the fact that it is easily digested and can break down undigestible oils to improve digestion and help cure scurvy. If an 18th-century person read this finding, he would certainly not have believed that fruit juice contained that which the body lacked and that it could prevent scurvy.

And so it took another decade, until March 5, 1795, before the authorities introduced lime juice as the essential means of preventing scurvy. James Lind did not live to see it; he died less than eight months earlier, on July 13, 1794.

In its essence, his experiment is the perfect example of one of Murphy's Laws – the law of basic research – as given at the beginning of this story.

Story 3.3: The snob and the 7,000 cannons

On October 21, 1805 at precisely 11:45 am GMT, Admiral Lord Nelson ordered a signal to be hoisted on the flagship HMS *Victory*, requiring a total of 29 signal flags to spell out: "*England expects that every man will do his duty.*" It was the start of the largest naval engagement of the Napoleonic Wars and the largest, and definitely best-known, in the history of large sailing vessels. It took place near Cape Trafalgar (*Cabo de Trafalgar*) near Cádiz, Spain, with the three biggest naval powers of the early 19th century facing off against one another. It was the combined fleets of the French and Spanish navies against the fleet of His Majesty George III. The sole purpose of the French-Spanish alliance was to end England's longtime dominance of the sea. By gaining control of the sea, Napoleon could control world trade and, furthermore, he would be able to invade the British Isles and confirm his domination of Europe.

Admiral Pierre-Charles Villeneuve was commander of the allied forces, and the leader of His Majesty's fleet was probably the most famous navy captain in all of history – Lord Horatio Nelson.

The allied forces had the upper hand in terms of number of ships and sailors. Admiral Villeneuve commanded 33 large warships. At the time of the battle, three of them – the 136-gun *Santisima Trinidad* and two 112-gun ships, the *Príncipe de Asturias* and the *Santa Ana* – were the biggest ships of their kind in the world. The *Santisima Trinidad* had four gundecks manned by over 1,000 sailors. It was the biggest and heaviest ship built in the Age of Sail. Over 26,000 sailors served in the allied flotilla. While Admiral Nelson had fewer warships – "only" 27 – and fewer sailors, in the region of 18,500, he had more guns: 3,918 compared to the 3,168 on the allied ships.

On that day near Cape Trafalgar, 60 ships – floating fortresses – with 45,000 men on board and over 7,000 loaded cannons, came face to face. It was the greatest naval battle that had ever been seen.

Every schoolchild knows how the Battle of Trafalgar turned out. The Royal Navy crushed the enemy, and what the British warships could not do, the next day's storm finished off. Barely five of the allied ships were saved, with 4,400

French and Spanish sailors killed and over 2,500 wounded. Another 14,000 were captured. The British losses were considerably smaller, but the biggest tragedy was the death of the fleet commander, Admiral Nelson. He was the most famous of the 448 sailors who died in the battle for the British side.

With his victory at Trafalgar, Admiral Nelson went down in history and secured a prominent place in Fiddler's Green, the legendary afterlife for sailors. Great Britain secured its dominance of the seas for more than 100 years.

Numerous books analyzing the memorable battle have been written in the more than 210 years since it took place. Although many reasons have been given for the British victory, it is generally considered that the strategic and tactical brilliance of Admiral Nelson was the decisive factor. But several other factors also came into play, such as the fact that the British fleet had plenty of experienced naval officers. That could not be said of the French fleet, most of them having been lost in the Revolution. Such officers were aristocrats and had been either executed or exiled. Additionally, the British sailors were more experienced and more motivated than their counterparts on the other side of the battle line.

But there was another reason, one that justifies our mention of this historic battle in a story about vitamin C. The Admiralty had ordered that 30,000 gallons of fresh lemon juice be included in the daily rations of the British fleet. Admiral Nelson, who as a young captain had almost died of scurvy, had ordered an additional 20,000 gallons of lemon juice for his men about six months before the Battle of Trafalgar began. Not one of the more than 18,000 British sailors had to be relieved of duty for scurvy, which just 20 years earlier would have been unheard of. Scottish physician Gilbert Blane deserves credit for getting the British Admiralty to change its long-term negative stance on citrus as a cure for scurvy – and maybe thereby deserves part of the credit for the victory at Trafalgar.

Sir Gilbert Blane of Blanefield, first Baronet, fellow of the Royal Societies of Edinburgh, London, and Göttingen, correspondent fellow of the Imperial Academy of Sciences of St. Petersburg and the Royal Academy of Sciences in Paris, fellow of the Royal College of Physicians, physician in the courts of King George IV and William IV, was a snob. And like every snob, he prized his high position in the social hierarchy. His cool treatment of persons of a lower rank earned him the nickname *Chilblain* from those he was in charge of, while he was toadying toward his superiors.

But Dr. Gilbert was also the doctor who saw to significant changes in the care of the health of sailors. He successfully advocated several dietary and hygiene changes, the most oft recited of which is the introduction of citrus juice to the naval diet. He was instrumental in the practical erasure of scurvy from the list of the diseases of Royal Navy seamen.

Blane was born on August 29, 1749 in the small village of Blanefield in southern Scotland, and was accepted at Edinburgh University at the age of 14. He initially studied theology before later deciding to switch to medicine. While at university, he was elected president of the Students' Medical Society, which shows that even as a student he was no stranger to ambition and social talent. Blane was at ease in social situations from an early age, so he had no trouble joining the higher circles in Edinburgh. He was recommended by several people as private physician to Lord Holderness, who was known in London not only for his numerous illnesses, but also for his good taste and the right contacts.

His acquaintance with the lord, and his successful treatment of the lord's various health issues, opened the door for Blane within London's social circles. It is unsurprising, then, that he soon became the private physician to naval captain Sir George Rodney. It was under his command that Blane first sailed at Christmas in 1799; and being selfless, hard-working, and devoted, he soon became physician to the entire flotilla under Rodney's command.

He would receive regular reports from other ships' surgeons, leading him to believe that the high levels of sickness on the ships of the Royal Navy fleet could be greatly reduced by improving the hygiene and diet of the seamen. His own statistics created from those reports showed that of the 12,000 men serving in the Royal Navy, one-seventh suffered from disease, primarily scurvy. Only 60 seamen died as the direct result of battle. Blane sent his findings to the Admiralty along with his recommendations: better ventilation, cleanliness and dryness, lime juice for the seamen, soap and a supply of basic medicines – all of which could reduce the rate of sickness among crews and increase their fitness for battle. The response was that the Admiralty had found that lime juice was less effective compared to the officially sanctioned scurvy treatment, barley broth, and it was far more expensive. It is hard to imagine now, but in the second half of the 18th century, sweetened barley broth really was the official treatment for scurvy approved by the Royal Navy, despite the findings of James Lind and others before and after him who described the positive anti-scurvy properties of citrus juice.

While he was unable to sell his truth to the conservative lords of the Admiralty, Blane at least convinced his commander of the need for changes. Soon after his recommendations were implemented, the fleet sailing under the command of Sir George Rodney saw its death rate reduced from one in seven seamen to one in 20.

Despite his efforts being ignored by the Royal Navy, Blane's reputation grew, as did his social status and number of useful acquaintances. When the spot of Commissioner of the Sick and Hurt Board opened up, it was taken up by Gilbert Blane. He was finally able to introduce the reforms he had long been recommending and which had proven successful during his time as physician to the flotilla of George Rodney. On March 5, 1795, a year after the death of James Lind and 49 years after his experiment aboard the *Salisbury*, lime juice was officially approved as the preventative for scurvy in the British Royal Navy. The daily ration was about four teaspoons, which is less than the recommended dosage today (more about that in the next story), but it was enough to keep the sailors healthy on their voyages. It was a general regulation of the Admiralty and soon people started to refer to sailors in the Royal Navy as "limeys."

British bureaucracy was slow and cumbersome, but it was also thorough. As soon as citrus juice was introduced into the seamen's diet, consumption grew and was soon at 200,000 liters annually.

On the day of the Battle of Trafalgar in October 1805, nine and a half years after lime juice became a mandatory part of the diet of the Royal Navy, each and every one of the 18,500 sailors under Nelson's command was perfectly fit. They were able to do all the hard physical labor involved in manning a ship and guns. It does not take much imagination to picture how the battle would have turned out if a third of the seamen had been so weak as to be barely able to stand on their own legs.

Sir Gilbert Blane died on June 26, 1834 at the age of 85. Many honors were bestowed upon him during his life, which he accepted as a matter of course, aware of his contribution to the health reforms that substantially improved the lives of his contemporaries. He was not a scientist, and he did not invent anything new, but with his strength of character and with the help of his social talents, his ability to appeal to the right people at the right time, he was able to implement reforms that ultimately earned him the title of "Father of Naval Medical Science."

Story 3.4: The Norwegian hygienist and guinea pigs

Admiral Nelson won the greatest naval battle near Trafalgar, but the battle with scurvy was far from over. Fortunately, the British naval flotilla had a solution. Sailors on ships from other countries, whether warships or merchant ships, continued to suffer and die from scurvy, but the condition was not limited to those at sea. Scurvy frequently afflicted people on the land too. In the American Civil War (1861–1865), more than 100 years after Lind's Treatise was published, 10,000 Union Army soldiers died as the direct result of scurvy. Another 45,000 died of the dysentery and diarrhea that followed scurvy and 10,000 men also died of scurvy during the California Gold Rush. A million people died in the Great Irish Famine in the 1840s, a large part of the deaths due to scurvy. Scurvy even became a pediatric problem after powdered milk began to replace mother's milk, because powdered milk did not contain any vitamin C.

It is inconceivable how many people died needlessly when a simple cure was already known, but the medical mainstream still did not consider scurvy to be a disease caused by a lack of vitamins. There were countless hypotheses about the cause of scurvy. One claimed that it was caused by putrefaction and could be cured by eating fermented foods. Another claimed it was caused by a lack of protein in the diet. Yet others claimed that the cause was contaminated food, infection, or lack of potassium and iron in the diet. In the early 20th century, there was even an official hypothesis that scurvy was caused by constipation and therefore could be cured with laxatives.

We now know that scurvy is caused by avitaminosis, a deficiency of vitamin C (ascorbic acid). Vitamin C helps in various enzymatic reactions, the most important of which is the synthesis of collagen, the most abundant protein found in mammals. It makes up a substantial part of the connective tissue – muscles and tendons – and is also present on the walls of capillaries. A deficiency brings about the typical symptoms of scurvy: brown splotches on the skin, swollen gums, bleeding, depression, festering wounds, and tooth loss. As mentioned, it is fatal if left untreated.

Humans are unfortunately among the few animals that cannot synthesize vitamin C. This an ability we lost during evolution approximately 60 million years ago. We are not unique in this, however. The same applies to some birds and fish, some species of bat, guinea pigs (remember this), exotic capybaras, and some primates. Since our bodies cannot produce vitamin C, we have to get it from food. There are various opinions as to daily amounts, with recommendations ranging from 40 mg to 2 g per day. That is far less than can be produced by those mammals who did not lose the ability to synthesize their own ascorbic acid, and 20–80 times less than is needed in the food of mammals who, like us, lost that ability. There is a large group of scientists who claim that we are actually underdosing on vitamin C.

We get vitamin C mainly from fresh fruits and vegetables. The highest content is found in the Australian Kakadu plum *(Terminalia ferdinandiana)*, for which ascorbic acid forms nearly five percent of its weight. Of the more readily available plants, rose hips have the highest vitamin C content – almost ten times more than oranges and lemons. For vegetables, the highest content can be found in sweet peppers (four times more than oranges) and broccoli (twice the amount contained in oranges). Apples and pears have a relatively low content (a tenth in comparison to oranges) and raisins and figs contain practically none. On the other hand, organ meats include a relatively high content of vitamin C. Raw calf's liver contains just slightly less per unit of weight than oranges, and fried chicken liver contains the same amount of vitamin C as grapefruit. The high vitamin C content in organ meats explains why the Inuit living above the Arctic Circle are not affected by scurvy, despite the fact that they have practically no access to plant sources of vitamin C (not in the past, at least).

Professor Axel Holst was not yet aware of any of this. His story illustrates the two principles of research. The first is that scientific research, and not only medical, is not linear. Often you must go back several decades to make a connection to some knowledge or other that has languished in the meantime. The second principle is that sometimes you need a little luck.

Axel Holst was born on September 6, 1860 in Christiania (now Oslo) into a family with many generations of physicians, so it was expected he would follow in the family tradition. The young Axel fulfilled the expectation with unprecedented exuberance. In 1893, at the age of 33, he was appointed Professor of Hygiene and Bacteriology at the University of Christiania and five years later, he was elected chairman of the Norwegian Medical Society. In addition to his professorial duties, Holst had a substantial impact on practical hygiene in Norway. He was the Medical Officer of Health in Oslo and a member of the Municipal Board of Health. He advocated for improvement of the poor hygiene conditions in working-class neighborhoods and brought attention to the pollution in the harbor. His opinions can still be considered liberal today, because he also unsuccessfully advocated for the regulation of prostitution. As chairman of a ministerial and parliamentary commission for alcoholism, he was strictly against prohibition.

The primary focus of his experimental work was nutrition. Holst studied the tropical disease beriberi and what he thought were similar diseases. At the time, the disease was known as "ship beri-beri" and in the second half of the 19th century, ship beri-beri was prevalent among Norwegian sailors. He even established a special commission in Norway to study ship beri-beri.

Early in his experiments, Holst noted the differences between the diseases with a similar name – the tropical one and the ship one. Unlike its counterpart, ship beri-beri responded positively to a diet containing fresh vegetables and potatoes. Today, we know how to explain that difference. Both diseases have one thing in common: they are caused by avitaminosis, i.e. a vitamin deficiency. Tropical beriberi is caused by a deficiency of vitamin B1 and mainly affects the nervous system. It occurs in areas where there is a deficiency of this vitamin, especially where the diet consists mainly of white rice which contains no vitamin B1. We can guess which vitamin was deficient in ship beri-beri.

At first, Holst used pigeons for his experiments, but was not satisfied and began looking for other animals on which to experiment. He chose guinea pigs. When he fed them the same food that caused symptoms of ship beri-beri in pigeons, the guinea pigs behaved differently from the pigeons. The symptoms displayed by the guinea pigs resembled those described in the literature as associated with a disease afflicting the human adult, namely scurvy. In his first essential paper, published in 1907 as a collaboration with his colleague Theodor Frölich, he devoted 20 pages to a detailed description of the differences between ship beri-beri and tropical beriberi, concluding that "*…facts speak in favor of the opinion that the described disease in guinea pigs is identical with human scurvy*." Their paper first used the term "ascorbutic" and associated it with the effects of a diet containing fresh vegetables. They also discovered that when cabbage is heated to a high temperature, it loses its ascorbutic properties.

The addendum to their paper is of interest to us. In it, the authors described the results of a similar experiment on dogs. As can be expected with current knowledge, the dogs did not show the same symptoms observed in the guinea pigs. They did not get scurvy because their bodies can produce vitamin C.

Some 160 years after the experiment aboard the *Salisbury*, the fact that scurvy was caused by a deficiency of something that could be found in fresh fruits and vegetables was scientifically accepted. Another 25 years would pass before it was discovered what that "something" actually was.

Professor Axel Holst was a notable scientist, a great orator, and a sharp debater. He was a true luminary and rightfully one of the big names in medicine in the late 19th and early 20th centuries. Today, though, we are left with the interesting question of what his life would have looked like and how our knowledge of vitamin C would have developed had they not opted to use guinea pigs as their laboratory animals rather than dogs, mice, or rats. Unlike guinea pigs, all of the aforementioned animals can produce their own vitamin C. They are not bothered by a lack of it in their food and do not contract scurvy.

Maybe we would still be treating scurvy with barley broth.

Story 3.5: The Hungarian politician and Hungarian paprika

In 1937, Albert Szent-Györgyi was awarded the Nobel Prize "*for his discoveries in connection with the biological combustion processes, with special reference to vitamin C and the catalysis of fumaric acid.*" A bit wordy, but all in all a fitting summary of his contribution to physiology and medicine. This award alone would be enough for him to have become a Hungarian hero and enter the pantheon of the greatest Hungarians, but his life was far more colorful and his scientific work and results were only a part of that life. Probably not even the most important part.

He was born on September 16, 1893 as Albert Szent-Györgyi de Nagyrápolt (written in reverse in his native language, i.e. Nagyrápolti Szent-Györgyi Albert), to a family whose roots went back to 1608. His grandfather and uncle were important physicians, both anatomy professors at the university in Budapest. He graduated high school with honors and in 1911 he entered medical school in Budapest. He said of himself that when he was young he was fairly lazy, so he dropped out of this school in short order. Albert was more drawn to his uncle's anatomy laboratory, and was allowed to work there under one crazy condition: the young man was to focus exclusively on the anatomy of the rectum and anus. Szent-Györgyi later wrote that, "*I started science at the wrong end.*"

Soon after the young Albert started science, World War I broke out. His studies interrupted, he was recruited as a medic into the Austro-Hungarian army. He served on the Italian front and fared quite well, earning a medal of valor, but after two years he had had enough. He carefully shot himself in the left humerus and pretended to be wounded by enemy fire. Albert got what he wanted, and after being sent home from the war he returned to finish medical school and married the postmaster general's daughter. Turbulent times awaited central Europe after World

War I ended, and they did not bypass Szent-Györgyi. He briefly worked in a pharmacology institute in the city of Pozsony, 125 miles from Budapest. Following the dissolution of the Austro-Hungarian Empire, Pozsony became a part of Czechoslovakia and its name was changed to Bratislava. As a Hungarian, Szent-Györgyi was forced to leave the city hastily. He began his travels to various European universities, but with his modest income he was unable to take along his wife and young daughter, so he usually traveled alone. From Bratislava, he went to a German university in Prague, then to Berlin, Hamburg, then Leiden and Groningen. He was in London briefly, then returned to Groningen in 1926. At each of these stops, he learned something new and was prolific in submitting articles to journals. British biochemist Sir Frederick Gowland Hopkins took note of some of his articles and arranged a Rockefeller Fellowship for him at Cambridge University. Albert Szent-Györgyi was finally financially secure, but truly at the last minute. He recalled that the loneliness and the separation from his family, coupled with the lack of money, had made him consider suicide at one point.

His work at Cambridge made him well-known as a scientist, and in 1931 the Hungarian Minister of Education offered him a job as head of the medical chemistry department at the University of Szeged. He had been a proud Hungarian his entire life, and he took the job without hesitation. He soon became known at the university not just for his fascinating lectures, but also for his unconventional leadership of the department. The university's board was not too thrilled by this, but it made him even more popular with the students.

Albert Szent-Györgyi was definitely not a conventional person. As a war hero decorated with the medal of honor, he could easily have waited out World War I without shooting himself. When the war ended, he would easily have been able find work as a general practitioner in Budapest and would not have to roam penniless around Europe. In the 1930s, he had a great job as head of a university department and he was a famous and respected scientist. But as soon as the far right government with fascist opinions came to power in Hungary, Szent-Györgyi was one of the first to stand up against the growing antisemitism and militarism. He actively supported his Jewish colleagues, and when there was nothing else he could do, he helped them to get out of the country.

When he received the Nobel Prize in 1937, he immediately became the hero of Hungary. When the Soviet Union invaded Finland two years later, he donated all of his prize money to the Hungarian volunteers fighting on the side of Finland. He protested the Hungarian alliance with Germany and had an active role in the Hungarian Resistance Movement.

In 1943, a group of Hungarian intellectuals and politicians planned to cooperate in secret with the allies, and Albert Szent-Györgyi was the go-between sent to negotiate the alliance. The meeting was to take place in Cairo, and Szent-Györgyi went there on the pretext of giving a science lecture. The meeting never took place. The Germans uncovered the plot and Szent-Györgyi had to spend the rest of the war running from the Gestapo.

The war ended and Albert Szent-Györgyi, as an active anti-fascist and world-renowned scientist, became a respected citizen and highly regarded figure. He was elected to the reconstructed Hungarian parliament and co-founded the Academy of Sciences. He was even considered as the potential president of post-war democratic Hungary. However, developments in Europe, especially the growing influence of the Soviet Union in its center, were simply not compatible with Szent-Györgyi's democratic principles. Disgusted, he emigrated to the U.S. in 1947.

Photographs of a young Albert Szent-Györgyi show him as a handsome man. He played piano and was good at many sports. He was married four times. He divorced his first wife, Cornelia, in 1941 and was married to his second wife, Marta, until her death in 1963. At the age of 72, he married for a third time, to the 25-year-old daughter of a colleague. The marriage lasted three years. He gave it another try at the age of 82, and his final marriage lasted until his death in 1986.

The isolation of the "antiscorbutic factor" had long eluded the efforts of scientists. The problem was that, unbeknown at the time, it was structurally similar to sugars, a lot of which can be found in fruits and vegetables. It was not easy to isolate one substance among many other similar ones. While he was in Cambridge, Albert Szent-Györgyi tried it with a different source – beef adrenal glands. They contain less vitamin C than fruit, but substantially fewer sugars. He found the substance he was looking for and was able to isolate a small amount of it.

When it came time to publishing his discovery, the publisher wanted the new compound to have a name. The unconventional Szent-Györgyi first suggested "ignose" (from the Latin "*ignosco*," meaning *I don't know*, and the suffix "*ose*" indicating a sugar). The editor rejected that name, so he tried another: "*Godnose.*" In the end, he had to settle for the name proposed by the editor – which is how the first publication about hexuronic acid came about. Preliminary analyses had shown that Szent-Györgyi's compound had six carbon atoms (*hex*) and was acidic (acid).

However, there was not enough of this hexuronic acid to analyze its exact structure. The antiscorbutic factor remained unknown. They did not even verify that it was in fact Albert Szent-Györgyi's hexuronic acid. The first headway was made in 1931, when Szent-Györgyi sent what remained of the substance to the University of Pittsburgh. It was confirmed in the laboratories there that hexuronic acid was indeed the substance they were seeking. The package he sent to Pittsburgh completely exhausted Szent-Györgyi's supply of the acid, and since he had no means of obtaining further beef adrenals, he had to look for another source. He did not have to look far. At the time, he was living in Szeged – the mecca of Hungarian paprika – and he attempted to obtain hexuronic acid from paprika. As we know from the previous story, sweet peppers have a high vitamin C content, so he was able to obtain nearly one hundred grams of pure crystals in a very short time. He finally had enough of the substance to analyze its structure and to confirm its effects. He also decided to change its name, inserting its main significance – the a(nti)-scorbutic properties. Ignose, Godnose, and hexuronic acid finally became what it is today – ascorbic acid.

It would be some time before the legitimacy of the name ascorbic acid was verified, but the Nobel Prize in 1937 was proof that the antiscorbutic factor had finally been discovered.

Ascorbic acid was not the only result of the work of Albert Szent-Györgyi. He lived to the age of 93, and influenced many areas of biochemistry, physiology, and medicine. He spent the final years of his research studying cancer. He died of kidney and heart failure resulting from leukemia.

In Hungary, he is still considered one of the most important people in the history of the country. In 2012, to commemorate the 75th anniversary of the award of his Nobel Prize, Hungary issued a silver coin with a value of 3,000 forints bearing the image of Albert Szent-Györgyi.

Story 3.6: The hardworking chemist and the role of wine flies

On Tuesday, October 18, 1905, a fanaticized throng hit the streets of Kiev. There were various groups in the crowd: monarchists, reactionaries, anti-Semites, and common criminals. Once in a while, a police or military uniform would flash among the crowd, which consisted mostly of Kiev residents. They had one single common goal: liberate Russia from the Jews. Immediately, if possible. The great Kiev pogrom had just begun.

The great Kiev pogrom of October 1905 was not the first or the last to be suffered by the Jews. At the same time, there were pogroms in over 600 cities and towns across today's Ukraine. Over the course of these pogroms, several thousand Jews were killed and their property destroyed. The one in Kiev in late October 1905 lasted around three days and saw the slaughter of 100 Jews, with a large portion of their property pillaged.

Young Tadeusz Reichstein was only eight years old at the time, but he remembered the bloody scenes of those days until the day he died. It was those memories that formed his entire, long life of almost 100 years.

He was born on July 20, 1897 as the fifth son of Jewish sugar industry engineer Isidor Reichstein. The Reichstein family had not lived in Kiev for very long, having arrived recently due to the work of the head of the family. However, after the October pogrom, Isidor Reichstein did not hesitate for a minute and took his entire family out of the city, emigrating to Switzerland. He probably saved all of their lives by doing so. His decision to leave was also of historical significance. If he had remained in czarist (and later Soviet) Russia, his son would not have invented the process for the synthesis of ascorbic acid or been awarded the Nobel Prize for his work.

Tadeusz's childhood alternated between good and not so good times. The good times included staying with his pharmacist uncle, where he first gained experience with chemistry. Or when, at the age of eight, he built a rudimentary laboratory in his parents' bedroom where he attempted to turn iron into gold in various ways. He was not successful. The bad times included boarding school, where he faced strict Prussian discipline in which punishment was meted out for every minor infraction. But the toughest were the years of World War I, when the family lost all of its assets. His father became seriously ill and his mother had to provide for the entire family.

Despite the hardships, Reichstein earned honors in school. Driven by a desire to see his family financially secure, he looked for a well-paid job right after he finished school. He found work in a small factory, where his job was to improve batteries for flashlights. He improved the batteries and used the money he earned to boost the family budget significantly.

Having secured the family's finances, Reichstein was able to begin his doctorate studies in 1921 at the Swiss Federal Institute of Technology *(Eidgenössische Technische Hochschule)* in Zurich. The theme of his work, the isolation of the volatile flavor of roasted coffee, was not very commonplace at the time. Nor was it easy, as the flavor and aroma of coffee are the result of a large number of various compounds working together. The project took him nine long years, but he ultimately succeeded. He and his colleagues identified, isolated, and patented a mixture of the basic compounds – over 50 of them – that together comprise the aroma of coffee.

When the project was complete, he decided to do something a little more serious and useful than isolate the aroma components of coffee. Since he had become a full professor in the meantime, he had his choice of fields to specialize in and chose the chemistry of vitamins.

Ever since the identity and structure of Albert Szent-Györgyi's ascorbic acid had been verified, a competition had begun to see who would be first to make it chemically. Ascorbic acid was a substance that had great scientific and commercial potential, and the person who devised a way to produce it synthetically would be rich and famous. But synthesizing ascorbic acid was not a simple matter. It consisted of several steps, and a necessary intermediate product was one specific sugar, sorbose. Sorbose was a known substance. Its structure was known, as was its role in the entire multistep process of synthesis. Obtaining it was a different matter. One way was to use specific bacteria that could transform sorbitol, a much more readily available sugar, into the desired sorbose. It was known that the type of bacteria required could be found in the slimy sediment that sometimes forms in wine vinegar or unpasteurized beer. Tadeusz Reichstein therefore decided to use that particular bacteria to prepare the sorbose. It was the right decision.

However, it was not easy and it would take many unsuccessful attempts. Finally, as with many scientists before him, he was helped by chance. In one of the experiments, he mixed sorbitol, yeast, and wine vinegar and left the specimens outside for several days. To his surprise, some of the specimens contained a deposit of white crystals of the anticipated sorbose. Moreover, a drowned wine fly was floating in one of the specimen jars and several crystals were attached to one of its legs. It was later determined that it was the wine flies which spread the type of bacteria (later named *Acetobacter suboxydans*) that Tadeusz Reichstein had so desperately needed. It took no time at all for he and his team to produce enough Acetobacter to begin large-scale production. The method became known as the Reichstein process and it is still used to produce ascorbic acid. He was one of the first to combine a chemical and biological method of production. The method is used to this day in modern biotechnology.

The history of vitamin C more or less ends here, and this could also be the end of this chapter, but we still owe something to our last protagonist, Tadeusz Reichstein.

He was only 36 years old when he devised his method for the chemical synthesis of ascorbic acid and he still had another 63 years of life ahead of him. Just like with the roasted coffee, when he finished his work on the synthesis of ascorbic acid, he finished that chapter of his life and got involved in something completely different – isolating hormones. After successfully improving batteries for flashlights and isolating the aroma of roasted coffee and the synthesis of ascorbic acid, his work in this field was successfully completed as well. Reichstein was able to identify 30 various compounds, the most important of which was cortisone, a substance that would later have an impact on numerous areas of medicine. The results he achieved in the chemistry of corticosteroids – which is what these hormones are called – were so significant that he was awarded a Nobel Prize for his work in 1950.

After retiring in 1967, Reichstein continued in his active life and, once again, he completed what he had set out to do. He decided to end his career as an organic chemist at the age of 70. He took a great interest in two of his hobbies, ferns and butterflies, and was successful once again. His intense study of ferns resulted in over 100 scientific papers on the subject and he became a highly respected luminary in this field, much as he had been in the field of organic chemistry. Maybe, somewhere on the other side of the world, a story is being created about a notable pteridologist who was once an organic chemist before he became famous for the science of ferns.

The next area Reichstein took up was also completely different from the others. He isolated substances from butterflies that were not found anywhere else. For instance, he was able to isolate and describe 30 brand-new compounds from the monarch butterfly (*Danaus plexippus*).

At first glance, it seems as though Tadeusz Reichstein was a lucky man – whatever he put his mind to, he achieved success – but his industriousness and talent contributed a great deal to his luck.

One last thing must be mentioned about Tadeusz Reichstein. His friends and colleagues unanimously considered him to be an incredibly generous and selfless man, someone who spent his entire life helping others. He paid for his younger colleagues' studies out of his own pocket; he established and financed orphanages; and he supported everyone he could as far as his means allowed. His character was on full display in the decision he made when he was accepting the Nobel Prize. It was awarded to him together with his two American rivals, and he expressed his gratitude in his speech, saying he owed his Nobel Prize to them for their work and contribution.

Concluding remarks

In our society today, with plentiful food of every kind, it is very hard to imagine that vitamin C was once a medicine, one that prevented disease and the death of many people. Modern methods of refrigeration and preservation allow a high content of vitamin C to be retained in foods, and the globalization of trade we have witnessed over the past years means you can have fresh fruit from any part of the world at any time you want. And if you still feel you have a deficiency, thanks to Tadeusz Reichstein you can take inexpensive vitamin C in tablet form. The current consumption is in excess of 100,000 tons annually.

Ascorbic acid is a frequent food additive for both nutrition and preservation purposes. Its antioxidant properties make it the most usable (and the safest) of the food preservatives. Ascorbic acid and its salts and compounds have been given the E numbers E300 to E304.

Those of us who live in the more developed countries are fortunate. Scurvy is far from being eradicated in the other, less fortunate parts of the world. It is a disease of deficiency for which no vaccine is available. It still occurs in arid regions and is a frequent guest in migrant camps, in times of war, and during natural disasters. It is everywhere there are shortages of food. Where there is hunger, there is scurvy. Fortunately, a lack of vitamin C no longer sets the course of history as it did in the past.

There are many lessons to be learned from the history of vitamin C. It taught us that the path to knowledge is not linear and that it is often necessary to go back and verify what was already known. It can also sometimes be better to give up scientific concepts and hypotheses and return to the knowledge of earlier generations.

4

Insulin

One of the timeless dreams of humanity is the discovery of a fountain of youth; an elixir of life; a beverage that takes away pain and death. The first famous "seeker" was Qin Shi Huang, founder of the Qin dynasty. He lived in the third century BCE in what is now China. In his old age, he was so terrified of death that he sent his court sorcerer, Xu Fu, to find an elixir from the mystical eastern mountains. Xu Fu was outfitted generously, with 60 boats, 2,000 seamen, and 3,000 young boys and girls. When they failed to find the elixir, they did not return and instead sailed on to Japan, becoming the first known settlers of the Japanese islands. At around the same time that the Chinese emperor Qin was looking for the elixir, we encounter it in ancient Greece as well. It was described by the philosopher Zosimos of Panopolis under the name Cheirokmeta. He was far from the only one; the elixir of life, along with the philosopher's stone, is an inseparable part of ancient philosophy.

Arab alchemists brought teachings about the elixir of life to medieval Europe, and the attempt to discover it was one of the main objectives of alchemy for many long centuries. But with growing knowledge, the search for the elixir became unscientific and gradually faded away. The elixir of life is still alive in fables and myths, and maybe even a little in our subconscious minds. There are few who look forward to growing old and dying.

In 1922, the elixir of life became a reality, at least for a specific group of people. This group was made up of diabetic patients and the elixir of life was insulin. Without it, they were condemned to die and there was no way to help them. The survival of patients with type 1 diabetes, mostly young people, was limited to just a few short years. With insulin, even if it had to be administered on a daily basis,

V. Marko, *From Aspirin to Viagra*, Springer Praxis Books,
https://doi.org/10.1007/978-3-030-44286-6_4

they could look forward to practically the same lifespan as healthy people. The discovery of insulin is one of the most dramatic events in the history of medicine.

Story 4.1: The bold experimenter and sweet urine

An unusual wedding took place on July 25, 1784 at an Anglican church in the city of Bath in the southwest of England. The bride was one Hester Lynch Thrale. At 44, she was not the youngest ever bride and this was her second wedding. Her first husband, wealthy brewer and member of the British parliament Henry Thrale, had died of a stroke a few years before, apparently caused by his insatiable appetite for food and ale. Although it was said there was no great love in their marriage, the couple had 12 children.

Hester Thrale was not only a rich widow, she was a refined and educated woman. Her husband's financial status allowed her to enter London society, where she mingled with the intellectual elite. She was a long-time friend and confidant of Samuel Johnson, one of England's greatest writers. Simply put, she was a beautiful, rich, sophisticated woman, and a member of the Anglican church.

The groom was the same age as the bride, one Gabriel Mario Piozzi, a singer, composer and music master to Hester's daughter, Queeney. He was a charming musician, an even more charming companion, and a Roman Catholic. For Hester Thrale, it was love at first sight and after a bit of wavering, they had their wedding ceremony. Well, two ceremonies: first, a Roman Catholic ceremony, and two days later, in Bath, a Protestant ceremony.

Their marriage caused quite a scandal in British social circles. Marrying a foreigner, a poor singer – and a Catholic of all things – was practically unforgivable for a lady of the English upper crust. The London newspapers were full of condemnation and nearly all of Hester Thrale's friends deserted her. Even her oldest daughter Queeney, the one taught to sing by Gabriel Piozzi, refused to accept him as her new father and left home. The only one who did not denounce the marriage – and the one who actually played a role in it – was her physician and sometime confidant, Matthew Dobson. He was the only one to wish her happiness in the face of society's judgment, and he was the one who persuaded her daughter to return home to her mother.

The wedding of Hester Thrale and Gabriel Piozzi has nothing at all to do with this chapter, but it does provide some insight into the protagonist of this story: Dr. Matthew Dobson.

Matthew Dobson was not just a good and kind person, he was also a person with a natural talent for philosophy, a bold experimenter, and a seasoned observer. He was born in 1732 at Lydgate in northern England as the son and grandson of Anglican ministers. Refusing to carry on in the family tradition, he decided to study medicine. He graduated from Edinburgh University, moved to Liverpool, and in 1770 was appointed physician at the Liverpool Royal Infirmary. He was one of only four doctors in Liverpool at the time.

Dobson was a daring experimenter and the range of his experiments is extensive. He attempted to treat scurvy and bladder stones with carbon dioxide and he researched the therapeutic effect of copper salts and the warm springs in the spas of Matlock, not far from Manchester. In 1775, he published an article about the effects of heat on human health. Dobson performed the experiment on himself and one of his colleagues, who shut themselves into a room that was about nine square feet and raised the temperature to 224 °F. They determined that their pulses were raised to 120 beats per minute, but their body temperature did not exceed 99 °F. They stayed in the heat for ten minutes, the heat cooked three eggs and they successfully ended the experiment. He wrote an article about the experiment entitled *Experiments in a heated room*, published in the magazine *Philosophical Transactions of the Royal Society* in 1775.

But it was his experiments with diabetes that were best known. It all began on Wednesday, October 22, 1772, when Dobson admitted a 33-year old man named Peter Dickonson to the hospital in Liverpool. Dickonson had exhibited the typical symptoms of diabetes for nearly eight months, with severe thirst, weight loss, fever, and excessive urination. According to Dobson's records, the patient was passing 28 pints of urine every day. That number was probably slightly exaggerated, but definitely pointed to a serious disorder. Dobson conducted a series of experiments, one of which he recorded as follows: "*Two quarts of this urine were, by a gentle heat, evaporated to dryness... There remained after the evaporation, a white cake which...was granulated, and broke easily between the fingers; it smelled like brown sugar, neither could it, by the taste, be distinguished from sugar, except that the sweetness left a slight sense of coolness on the palate.*" While this urine-tasting may seem a bit unappetizing, this is a record of a classic diagnostic procedure that was considered standard in Matthew Dobson's times. Dobson went on to observe that the blood serum was also sweet, although not as sweet as the urine. He was really the first to describe hyperglycemia as a part of the disease.

Matthew Dobson was not the first to associate the disease known today as diabetes mellitus with excessive secretion of sugar in the urine. The disease was recognized in ancient times, with the first written mention in the *Ebers Papyrus* (we have already encountered this a few times in our earlier stories), which also described the treatment. It consisted of a decoction of bones, wheat grains, bran, lead and earth. The papyrus is silent as to the effects of this treatment. The term diabetes comes from the ancient Greek word meaning, roughly, "to pass through," so it actually describes one of the primary symptoms of the disease, which is excessive urination. It first appeared in Greek texts around 2 BCE. Centuries later, the Greek physician Aretaeus of Cappadocia described the symptoms in detail: "*...the patients never stop making water, but the flow is incessant, as if from the* *opening of aqueducts...but the patient is short-lived...life is disgusting and painful; thirst, unquenchable; excessive drinking...is disproportionate to the large quantity of urine ...one cannot stop them either from drinking or making water. Or if for a time they abstain from drinking, their mouth becomes parched and their body dry; the viscera seems as if scorched up; they are affected with nausea, restlessness, and a burning thirst; and at no distant term they expire.*" It is interesting that Aretaeus, just like Galen, another famous physician, described the disease as very serious but also very rare.

Physicians in ancient India also made the association between excessive urination and the sweetness of the urine. They discovered that the urine of diabetic patients was sticky, tasted like honey, and attracted ants. There are primitive peoples today still using this method to diagnose diabetes. Patients with suspected diabetes urinate on an anthill and physicians observe the behavior of the ants. The sweet urine symptom of diabetes is also evidenced by the Chinese name for diabetes, *táng niǎo bìng*, which translates to "sugar-urine disease." But we do not have to look to foreign languages; for years, the disease was known in English as "sugar diabetes."

In the fifth century AD, the Indian physicians Sushruta and Charaka first described the difference between the two types of diabetes. The young and impoverished were most often afflicted with the first type and the prognosis was poor. The second type was typical for older and overweight patients and had a substantially longer survival rate.

In later years there were increasing mentions of the disease, but the discoveries of those authors did not add much to the knowledge that had been gained by authors in ancient times. Perhaps the only exception is Thomas Willis, one of the greatest English physicians of the 17th century. A work that was published after his death, entitled *Diabetes or the Pissing Evil*, contains sentences that are worth citing even today. He wrote that in the past, diabetes had been rare, "*...but in our age, given to good fellowship and gusling down chiefly of unallayed wine, we meet with examples and instances enough, I may say daily...*".

But let us return to Matthew Dobson's patient, Peter Dickonson. He spent several months under the care of his physician and Dobson tried a variety of treatments on him, including a rhubarb or senna decoction, mixtures containing opium, or Spanish fly. He even offered to pay the expenses for a spa treatment, but Dickonson never went to the spa, apparently refusing to continue with the treatments ordered by Dr. Dobson. The fate of the patient is unknown.

Little is likewise known about Dobson's personal life. We know that he married Susannah Dawson at the age of 27. She was an educated woman who later became a famous translator from French. They had three children, but the marriage was not a happy one. While Matthew was kind and respectful to his patients, one of his contemporaries described his wife as "*coarse, low-bred, forward, self-sufficient... a strong masculine understanding*." She may have been a closet lesbian, which her relationships with women she knew appear to support.

We also do not know what Matthew Dobson looked like, since no portrait of him exists. There had been a portrait hanging in the Liverpool hospital, but in 1874 a local physician used it as a target for pea-shooting practice and it was severely damaged.

Matthew Dobson lived a short life. His poor health caused him to leave Liverpool in 1780, and he died four years later at the age of 49. His work did not have much of an impact during his lifetime, but 25 years later, it became the basis for the first treatment plan for diabetes – the diabetic diet.

Story 4.2: The military doctor in Barbados and various diets

Ancient Indians knew that diabetes patients have sugar in their urine, and this was confirmed by the English physician Matthew Dobson, the protagonist from our last story. The question was, how did the sugar get there? The prevailing hypothesis was that diabetes was caused by a malfunctioning of the kidneys. A symptom of the disease is excessive urination, and since the kidneys are responsible for excretion, then a kidney malfunction seemed logical. Dobson arrived at a different conclusion. When he discovered that it was not only the urine that was sweet, but the blood serum as well, he realized the malfunction must be somewhere else and that the kidneys were innocent.

Scottish military doctor John Rollo got his hands on Matthew Dobson's work 25 years later. He had also correctly determined that the disease is not caused by poorly functioning kidneys. By that time, it was already known that sugar enters the body primarily with a starchy diet and substantially less with a meat-based diet. Food is processed in the stomach, so the problem had to be a malfunction of the stomach, putting too much sugar into the body because of a starchy diet. By limiting the amount of carbohydrates in the diet, the problem would be solved. Thus, the first diet intended to help diabetic patients was created. It was also the first systemic approach to the management of a serious illness.

John Rollo's date of birth is unknown, but it must have been somewhere around 1750, because in 1776 he was already a physician in the Royal Artillery. From 1778, he served in the English colonies, in St. Lucia and Barbados. In 1794, he was promoted to surgeon-general. Upon returning home, he oversaw the construction of a new addition to the Royal Artillery Hospital at the Royal Military Academy. Rollo died on December 23, 1809. He was a well-rounded observer, writing several publications about the medical conditions in the garrisons where he worked.

Rollo encountered his first diabetic patient in 1777, before he shipped off for distant ports. The patient was a weaver from Edinburgh, although not many details have been preserved from this case because, as John Rollo wrote: "*I well remember that the blood and urine exhibited the appearances described by Dr. Dobson; but the papers, and a portion of the saccharine extract which I carried with me abroad, were lost in the hurricane at Barbados in 1789.*" The job of a military doctor in exotic countries was not all fun and games.

The records on his second patient are far more detailed, including not only information about the patient but also about the diet Rollo put him on. The patient was Captain Meredith, an acquaintance of Rollo from Barbados. When they met in 1794, the 32-year-old captain showed no signs of illness but, as the physician wrote, "*...he always had impressed me, from his being a large corpulent person, with the idea that he was not unlikely to fall into disease.*" Meredith paid a visit to Rollo on June 12, 1796, and the very first thing the doctor noticed was that he had lost a lot of weight and his face was ruddy. He complained of always being thirsty and hungry, and since he had to drink so often, he also had to urinate often. Diabetes immediately came to John Rollo's mind, and after finding sugar in the patient's urine the diagnosis was verified. He was surprised, however, that his colleagues had not made the same finding, since Captain Meredith had been having these problems for several months. Furthermore, in that time the patient's weight had fallen from 232 to 162 pounds, which was too low for a man of his height of nearly six feet.

The therapy ordered for Captain Meredith by John Rollo was truly complex. It consisted of not only a diet, but also another five components. It looked something like this:

Firstly, the diet...
Breakfast: One and a half pints of milk and half a pint of lime-water, mixed together; and bread and butter.
Noon: Plain blood-puddings, made of blood and suet only.
Dinner: Game, or old meats, which have been long kept; and as far as the stomach may bear, fat and rancid old meats, such as pork. To eat in moderation.
Supper: The same as breakfast.

Secondly.
A drachm of kali sulphuratum (mixture of potassium sulphide and potassium sulphite) to be dissolved in four quarts of water which has been boiled, and to be used for daily drink. No other article whatever, either eatable or drinkable, to be allowed, than what has been stated.

Thirdly.
The skin to be anointed with hog's lard every morning. Flannel to be worn next to the skin.

Fourthly.
A draught at bed-time of 20 drops of tartarized antimonial wine (used as an emetic) and 25 of tincture of opium; and the quantities to be gradually increased. In reserve, as substances diminishing action, tobacco and foxglove.

Fifthly.
An ulceration, about the size of half a crown, to be produced and maintained externally, and immediately opposite to each kidney.

Sixthly.
A pill of equal parts aloes and soap, to keep the bowels regularly open.

The daily diet contained only about 600 kilocalories of carbohydrates and 1,200 kilocalories of fat. The added components were completely standard for the time. It is surprising, however, that Rollo did not include bloodletting in the treatment, which in those days was the universal approach to treating all diseases. The ulceration opposite to each kidney was a method frequently used in the 18th century. It was intended to prevent the infestation and inflammation of the internal organs. The antimonial wine was introduced as a treatment by the French a century earlier.

Captain Meredith began the treatment on October 19, 1796, and two weeks later, "*he drank only three pints of water, and made only two quarts of urine, which…was not sweet.*" Rollo gradually simplified the therapy and took away some of the components (opium). However, he insisted that the basic rules be followed, mainly a meat diet and limited physical activity. After two months, the patient no longer felt thirsty and his urine output was normal. The treatment continued with a slightly less restrictive plan and the captain was allowed to eat a little more bread.

Doctor Rollo's diet was not completely unnatural. The diet of the Inuit living in the Arctic regions, or the shepherds living in the Pampas in South America, is constituted to a large extent of only animal products. The significance of Rollo's diet is that he was treating the disease rationally for the first time, trying to prevent the production of sugar. It did not take long for several others to follow suit. Many of the later diets were even more brutal than John Rollo's. For instance, the diet of Arthur Scott Donkin of Sunderland consisted of only skim milk and patients were not allowed to eat or drink anything else. It is not hard to imagine why few could tolerate this diet. The creator of the diet himself said that the diet could only be followed by patients "*…placed in isolated special wards, and under the care of strictly trustworthy nurses.*"

The German physician Carl H. von Noorden, who held positions at universities in Frankfurt and Vienna in the late 18th and early 19th centuries, created a diet consisting only of oatmeal. One of his adherents described it like this: "*250 g oatmeal, the same amount of butter and the whites of six or eight eggs constitute the day's food. The oatmeal is cooked for two hours, and the butter and albumin stirred in. It may be taken in four portions during the day. Coffee, tea, or whisky and water may be taken with it.*" Probably the least of the problems with this diet was figuring out what to do with all the extra egg yolks. Others in this line of diets include the potato diet and the fat diet. The names of those diets tell you everything you need to know.

There was one diet that was dramatically different from the others. It was invented by the eccentric French physician and poet, Pierre Piorry, who rejected all the existing procedures and began treating diabetes with sugar. His thought process was that the reason diabetics lose weight and feel so weak is that they lose large amounts of sugar in their urine, so replacing that sugar should restore their strength and health. As expected, the therapy did not bring any improvement and its popularity waned quickly.

All the diets, even the diets of today, require strict discipline and activity as well as the active engagement of the user. The more difficult the diet, the more active engagement is required – and that is its greatest downfall. The diets that were advocated for the treatment of diabetes were too stringent and were practically impossible to follow. Rather than suffer the restrictions of a diet, the patients took a pill, no matter how disgusting the pill was – and in the 19th century, there was a plethora of awful pills. According to official records, in 1894 in the U.S. there were 42 anti-diabetic medications available on the market. They contained everything you could imagine, such as bromides, uranium nitrate, and even arsenic. They were often dissolved in alcohol, which was probably the only ingredient that had any type of positive effect. Instead of alcohol, the fancier remedies used red wine from Bordeaux. These so-called patented medicines were for the most part shameless scams on patients. Nevertheless, many doctors praised the remedies and prescribed them to their patients, and the patients used them willingly, especially when the advertising made claims like, "*Dill's Diabetic Mixture, the only known remedy for the deadly disease. No dieting is necessary.*" Or, "*As soon as the patient has made use of this wine, his thirst is allayed almost instantaneously; his strength reappears; all his functions are gradually restored.*" As patients, it would seem we are little changed.

John Rollo's diet, just like the diets of his successors, was used but never enjoyed wide-spread popularity. These are all part of the history of diabetes treatment, if for no other reason than at least to illustrate how difficult and often hopeless the treatment was 200 years ago.

That remained the status quo for quite some time; about a century, to be exact.

Story 4.3: Two diabetologists, starvation, and Elizabeth the Iconic

There are many therapies in the history of medicine that today would be viewed as brutal and drastic, and would be absolutely unacceptable in our modern environment of humanitarian and ethical approaches to treatment. At the time these were invented, however, many were very widely used. There was no other choice, because often they presented the only possible remedy, one that could at least somewhat improve the condition and lives of suffering patients. Our stories delve into some of those therapies. One of them was the starvation of diabetic patients, introduced in the early 1920s.

In our first story about diabetes, we briefly mentioned that ancient Indians already recognized there were two types of diabetes. The first type mainly affected the young and the impoverished, and the second type mostly affected older people and obese patients. One example of a patient with the second type was the acquaintance of John Rollo, Captain Meredith, whom we met in the previous story. Fortunately, the majority of patients – roughly 90 percent – fall into the second group. Fortunate because this type of diabetes was and is easier to control and patients have a much better chance of survival. These sufferers were also the target group for the diet campaigns mentioned in the previous story.

Patients with type one diabetes had a much harder time of it and until the discovery of insulin, their disease was untreatable. Most people who were diagnosed with diabetes before they turned 20 years old died within two years after the first symptoms appeared.

In an effort to ease the bleak fate of patients with type one diabetes somewhat, two American diabetologists, Frederick Allen and Elliot Joslin, came up with a treatment they thought would be successful: starvation.

Frederick Madison Allen was the younger of the two, but he was the one who invented the therapy. Colleagues characterized him as a "*stern, cold, tireless scientist, utterly convinced of the validity of his approach,*" or as "*a terse man who never smiled and treated his patients like laboratory animals.*" He was a doctor and experimenter, who characterized the first years of his scientific career as "*living like a hermit, continually working seven days a week.*" He spent hours experimenting on laboratory animals in an effort to find the cause of diabetes and how it could be treated, paying the costs of most of the experiments out of his own pocket. After three years of intensive research, he had enough results to publish. His manuscript was so extensive that he could not find a publisher willing to wade through the entirety of it, so he borrowed $5,000 from his father and published the 1,179-page book under the title *Studies Concerning Glycosuria and Diabetes* at his own expense in 1913.

His basic diet for the first ten days consisted of 3–4 pints of water, 1–3 cups of coffee, 2.5 cups of clear meat soup, and 3–6 bran muffins. Unlike his predecessors in earlier centuries, Allen had the benefit of methods that allowed for the relatively simple and quick measurement of sugar concentration in the urine. When sugar was no longer present in the urine, proteins and fat were added to the diet until the daily calorie intake was deemed sufficient. The daily calorie intake of this diet was far less than 2,000 kilocalories, with most of them obtained from fat. Just to compare, today's recommended calorie intake for the average 20-year-old man is 2,500–3,000 daily. In many aspects, Allen's diet was not that different from the others. Where it did differ, however, was that it caused persistent malnutrition in his patients.

Allen's starvation diet did have positive results, especially since he typically presented the results on a case-by-case basis without any statistical evaluation. The truth of it is, in those days, doctors had little else available.

A great promoter of Allen's diet was his colleague and one of the greatest diabetologists of the 20th century, Elliot P. Joslin. He was so well-known and popular in medical circles that he was often referred to only as *EPJ*. Biographers have characterized him as "*a man of deep faith in religion and science, frugal but immaculate in dress, rising early, working long hours, eating sparingly.*" While Allen was mostly an experimenter, Joslin had a good bedside manner and since he was able to pepper his puritanical personality with charm and optimism, patients loved him.

He first encountered diabetes as a student when his aunt was diagnosed with the disease. Aunt Helen was also one of the first of nearly 1,000 patients whose progress he followed and conscientiously recorded into patient ledgers. Joslin's collection of 1,000 patient ledgers was the largest register of diabetic patients at the time. It was part of his textbook entitled *The Treatment of Diabetes Mellitus*, which was the first of its kind in the English language.

The same principle applies to the "starvation" diet as applies to the diets mentioned in the previous story: without discipline and the cooperation of the patient, it cannot work. There were doctors who complained that the patients were not dutiful and lacked the fortitude, causing them to withdraw from the treatment. The consequences were always fatal. Soon after they came off the starvation diet, they died. It was Joslin who actively worked on patient cooperation instead of criticizing. One of his famous quotes is: "*The person with diabetes needs character building just as much as body building…*" He published the first manual for doctor and patient that advised patients how to take control of the disease.

But even patients who were conscientious and found the fortitude did not have it easy. Elliot Joslin declared in his textbook that his therapy reduced the mortality rate by 20 percent, but an extended life was accompanied by suffering. Patients had two choices: either they could not withstand the starvation and died of diabetes soon after, or they continued the starvation and died of malnutrition. This was particularly true for younger patients, whose extended lives were really just a slow death by hunger. The fate of one ten-year-old boy serves as an example. He lived with diabetes for more than four years, but he only weighed 26 pounds. Pictures from the diabetes wards in the early 20th century bear a very strong resemblance to the photographs of what was discovered 40 years later when the Nazi concentration camps were liberated. But what could they do if nothing better was available?

Fortunately, starvation as a way of prolonging the life of diabetic patients did not stay around for very long, only about ten years. It ended with the discovery of insulin in 1922. In the history of insulin, a young patient named Elizabeth Hughes became an icon and her story perfectly characterizes the difference in the treatment of diabetes in the pre-insulin and post-insulin periods. The difference between the times when doctors were only able to describe the disease and monitor its course, and the new era when they started to have the ability to take an active role in controlling the course of the disease and save patients' lives.

Elizabeth was the daughter of Charles Evans Hughes, a prominent American politician in the early 20th century. She was born in 1907 and had a happy childhood in a well-to-do American family until she was diagnosed with diabetes at 11 years old. She came from a family of means, and so in the spring of 1919 she became a patient of the well-known Dr. Frederick Allen. From the very beginning he put her on his very strict diet, of 400 kilocalories a day. When the therapy started, she weighed 75 pounds, after two years she weighed 53 pounds, and a year later she weighed just 45 pounds. She did not have the strength to walk but just sat, read and did needlework. She was facing the same fate as most young diabetics, a slow death.

But Elizabeth was lucky. In 1922, she and many other patients got their miracle. In Toronto, Canada, after many failed experiments, a medication for treatment of diabetes – insulin – was successfully prepared. That same fall, the 15-year-old Miss Hughes became one of the first diabetes patients whose life was saved with insulin treatment.

The very first injection removed the sugar from her urine, and after a week Elizabeth was on a 1,200-kilocalorie-a-day diet, three times what she had been eating before. After another week, she was up to a healthy 2,200 kilocalories and for the first time in almost four years, she was allowed to eat bread and potatoes. She returned home towards the end of the year, and by January she weighed nearly 110 pounds.

Elizabeth grew up, went to college, and gave birth to three healthy children, eventually dying of a heart attack at the age of 74. She lived practically a normal life thanks to insulin, having received a total of 43,000 shots.

Insulin put an end to the drastic starvation diet era of Frederick Madison Allen and Elliot P. Joslin. Once the application of insulin became standard medical practice, the paths taken by these two great men were very different. Allen left diabetology and focused on other areas of medicine, but not very successfully.

Elliot Joslin continued to work in diabetology and worked his way to becoming one of, if not *the*, greatest diabetes experts of the 20th century. In 1952, his practice was renamed the Joslin Clinic. Today, the *Joslin Clinic* is located at *One Joslin Place* in Boston and is the largest diabetes treatment center in the world.

We will close this story with just one more observation about the destiny of young Elizabeth Hughes. In 1922, before insulin became the standard treatment for diabetes, Elliott Joslin asked himself the question that had been plaguing generations of diabetologists: "*If diabetes is always fatal in children, why prolong the agony? Why not just let the poor children eat and be happy for as long as they are alive?*" He knew the answer right away: "*Because nobody knows whether a cure will be found in a year or in a month.*" Had Elizabeth Hughes not undergone Frederick Allen's drastic diet that prolonged her life, she probably would not have lived long enough to be liberated by insulin.

Story 4.4: The vivid scientist from Mauritius and the elixir of youth

What causes diabetes? And why is insulin the wonder treatment? The answer to the first question is also the solution to the second.

In the early 19th century, the organ that was responsible for producing the symptoms of one disease or another was determined during autopsy. For example, if a person gradually became weak, was pale, and coughed up blood, an autopsy would show changes in the lungs typical for tuberculosis. But it did not work that way for diabetes. Not even the most exhaustive pathological and anatomical examination showed any changes in any of the internal organs. In spite of the polyuria, the kidneys of diabetic patients were no different from the kidneys of healthy patients. It was therefore held for many years that diabetes was a constitutional, or general disease that was not caused by any specific organ.

The first concrete focus on an organ that could be responsible for diabetes came in the mid-19th century. Claude Bernard, one of the greatest physiologists of the 19th century, discovered that the blood flowing from the liver into the body contains more sugar than the blood that flows from the digestive tract to the liver. Could the liver be responsible? Bernard thought so. He discovered a starchy substance in the liver and named it glycogen. Glycogen breaks down to form sugar, which is released from the liver into the body.

Another hypothesis was that the brain was responsible for diabetes, because animals did not have diabetes. The disease was perceived to afflict educated people predominantly, so those who made more use of their brains would be more prone to diabetes than uneducated people. We now know that this particular hypothesis did not take into account the role of physical activity as a strong factor of prevention.

A simple experiment conducted by another physiologist, Oskar Minkowski, provided the solution. During a discussion with a colleague, it occurred to him that diabetes could be related to the pancreas. The simplest way to verify this hypothesis was to remove the pancreas and observe what happened. He had plenty of laboratory animals available, so he got to work immediately. A few days later, one of the lab assistants complained that a dog that had previously been house-trained was now urinating everywhere. It was the dog from which Minkowski had removed the pancreas, and a test for sugar was positive.

The perpetrator had been discovered. The pancreas is the organ responsible for diabetes. The pancreas is an internal organ about 4.7–5.9 inches long, located in the abdomen near the liver and gall bladder. It has two functions, the first of which is the formation of pancreatic juice that is released into the small intestine to aid digestion. This is considered external secretion and is less important to our story. In this story, and for the entire chapter about insulin, we are far more concerned with the other function of the pancreas. In small areas of the pancreas that account for less than five percent of its total mass, a hormone is released through internal secretion; that hormone has the same name as this chapter – insulin.

The first knowledge that there are glands that produce, through internal secretion, substances that were later named hormones, came shortly before Oskar Minkowski conducted his experiments. The theory of internal secretions was first elaborated by Charles-Édouard Brown-Séquard, one of the most colorful scientists of the 19th century. This scientist is so interesting that he deserves a place in our story.

Imagine a tall, dark man with silver hair and an immaculately trimmed beard, not unlike Karl Marx, who was a year his junior. In contrast to the working-class movement ideologist, the dark eyes of this scientist reflected a colonial dignity, or at least that is how he was painted by an unknown artist when Charles-Édouard was 60 years old. He had been an outsider his entire life. He had a disdain for money, but worked tirelessly for as many as 20 hours a day. Brown-Séquard was not a scientist who stuck to one subject until it was thoroughly researched and documented, and instead jumped from one subject to another without concern for recording detailed evidence of new facts. He was satisfied at having discovered them. Over the course of his life, he published 577 papers and had an influence on neurology, endocrinology, and transplant medicine. Despite all this, he remained forgotten. The only reminder of this great man is a syndrome that causes spinal cord damage which was named after him – Brown-Séquard syndrome.

Charles-Édouard Brown-Séquard was born at 11:00 am on April 8, 1817 in the city of Port Louis, Mauritius. His father, Edward Brown, was an American sea captain and his mother was a Mauritian Creole with French roots. He never knew his father, who was killed by pirates before his son was born as he was returning to his home port with a ship full of rice. Charles-Édouard was raised by his mother, who earned money as a seamstress and by renting rooms to English officers travelling between Europe and India. He added her name to his as a second surname: Séquard.

Charles-Édouard got his wanderlust from his father. He crossed the Atlantic more than 60 times and spent nearly six years of his life at sea. He settled in America on four separate occasions, in France six times, once in England and twice in his native Mauritius.

He took his first ocean voyage at the age of 21, from Port Louis to Paris. He carried with him the plays, novels, philosophical essays and other literary works he had created and arrived in Paris with the intention of launching his career as a writer. His ambitions were grand, but the Parisian critics were not very appreciative of his talent, so he burned all his manuscripts in disgust and enrolled at medical school.

Even before finishing school, he decided he would not be a standard doctor. He was far too active, too restless, too hungry for new knowledge. He set his sights on experimental medicine right after he graduated, but paid work was scarce and he lived in poverty. Since he could not afford to rent a laboratory or buy laboratory animals, he borrowed lab space from his colleagues and sometimes procured laboratory animals by pilfering them from friends.

His next journey took him across the Atlantic. As a liberal-thinking young man, Charles-Édouard feared reprisals in monarchist France so he escaped to America, but he did not escape poverty. He earned money by giving French lessons and delivering babies, at $5 per birth. He married and returned to Paris a year later.

His entire life continued along the same vein. From Paris to his native Mauritius, where he was awarded a medal for his contribution toward ending a cholera epidemic, and then on to Richmond, Virginia. He did not remain in Virginia for very long, as he was not a proponent of slavery, and he returned to Paris. He founded a journal entitled *Journal de la Physiologie de l'Homme et des Animaux* (Journal of the Physiology of Man and Animals) and in the same year he moved to London. There, he helped establish the National Hospital for Nervous Diseases and became a highly regarded physician, corresponding with Charles Darwin and Louis Pasteur. He was also elected a fellow of the Royal Society of London.

Charles-Édouard then left London for Harvard University in America, and from there he went to Dublin, and then to Paris. And so on, and so on. He lectured, founded more journals and collected additional science honors. He experimented and published the results of the experiments. He had supporters as well as detractors, who called him the "*greatest torturer of animals*," but he did not limit his experiments to animals. When he wanted to learn about the processes that occur in the body after death, he used guillotined criminals, injecting them with his own blood.

Probably the most famous discovery of Charles-Édouard Brown-Séquard was organotherapy. The method itself had been known since antiquity and involves the use of parts of the organs of animals and humans for treatment purposes. The protagonist of this story lent new blood (not his own this time) to the ancient method. On June 1, 1889, the 70-year-old scientist presented his revolutionary findings to members of the *Société de Biologie* in Paris. It was really quite simple. He had been subcutaneously injecting himself daily over a two-week period with fluid extracted from the testicles of dogs and guinea pigs. The results were extraordinary, having improved his strength and stamina. He could lift heavier loads and walk faster up the stairs, and his urine stream was a full 25 percent more powerful.

The response was immediate, and the dramatic news drew the attention of not only medical circles but of the general public as well. The press picked it up and ran with it, and it is no wonder; the subject was quite sexy. The story had aging and rejuvenation and extracts from animal testicles, all presented as a strictly medical method and not some charlatanism. Newspaper headlines and magazine articles quickly appeared, calling it the elixir of youth and a panacea. The popularity of Charles-Édouard Brown-Séquard rose to dizzying heights. His portraits and caricatures appeared in newspapers and magazines on both sides of the Atlantic. Jokes and songs were written about him.

It goes without saying that the pharmaceutical industry took an interest in the discovery as well. Countless animal extracts became available bearing any number of variations of the word "elixir." The English company, C. Richter & Co., probably went the furthest. In the early 20th century, this company sold an extract from the testicles of rams and bulls, using the name of the discoverer in the product's name – Séquardine. Their advertisement claimed that the product could treat 17 conditions, including neurasthenia, anemia, kidney and liver disorders, podagra, sciatica, overall weakness, and the flu.

While Brown-Séquard enjoyed popularity with the general public, the discovery damaged his reputation within the professional sector. Although he had a few adherents, many of his contemporaries distanced themselves from him and considered him a quack. One of his detractors even said something to the effect of it being necessary to discuss not only the protection of the laboratory animals, but of the experimenters themselves. It was no help when he refused to accept any share of the earnings on the sale of the elixirs that were produced on the basis of his research. In one obituary, the last years of the research of Brown-Séquard were characterized as "*outright errors of senility.*"

Organotherapy, as interpreted by Charles-Édouard Brown-Séquard, could never work. When American researchers at the University of Chicago isolated testosterone in a dose that could cause the changes observed by Brown-Séquard 30 years later, it had required over 44 pounds of bull testicles. It was one of many errors that can be explained by the placebo effect, when a patient has so much faith in the success of a therapy that they can mobilize their inner strength. The positive outcome is then attributed to the (ineffective) therapy.

Even though organotherapy was a dead end, it helped to develop the science discipline that is known today as endocrinology – a branch of biology and medicine dealing with the endocrine system. Long before discovering organotherapy, it was Brown-Séquard who predicted that some glands secreted essential substances, the lack of which could cause health problems. At practically the same time as he wrote publications about the extracts from testicles, he also postulated on internal secretion in the pancreas, which plays a more important role in human life than the secretion of digestive juices.

Charles-Édouard Brown-Séquard – unorthodox scientist, tireless seeker, eccentric genius – died of a stroke on April 1, 1894. Characteristic of his nature, right up until his death he kept his friends updated on the progress of his own illness. He has been completely forgotten with time, and all that remains is the Brown-Séquard syndrome in neurology. It is unlikely that any aspiring physician knows who the syndrome was named after.

Just one final thought. It would appear that miracle elixirs that work on the principles of organotherapy have not been completely forgotten, they have just found new opportunities and a new target audience. A Japanese manufacturer of sports nutrition drinks is currently promoting something new for marathon runners: hornet juice.

Story 4.5: The aspiring amateur and the elixir of life

On Thursday, February 20, 1941, at around 8:30 pm, Captain Joseph Mackey saw the starboard engine of his Lockheed Hudson Mark III fail. At the time, he was flying east of Gander, Newfoundland, from where he had taken off about 30 minutes before. He was on his way to England, where his passenger was going to help test an anti-gravity suit. He reversed course to return to Gander, but then the port engine failed as well. The airplane crashed into the trees near a lake called Seven Mile Pond, 52° 46′ 14.3″ latitude and 56° 18′ 8.295″ longitude. William Bird, the navigator, and William Snailham, the radio operator, died in the crash. Captain Joseph Mackey was rescued. The sole passenger, Sir Frederick Banting, the principal discoverer of insulin and Nobel laureate, died the next day from his injuries. He was 49 years old.

If you were to fit the life story of Frederick Grant Banting into a few lines, it would read like something out of a Harlequin novel. It would be about a modest, timid and not very talented farmer's son from a small town in the Canadian province of Ontario, a boy whose own strength and resolve brought him glory in the world of science. It would also be a story about a man who was hellbent on achieving his goals; who studied medicine against his parents' wishes; who delved into solving problems that more experienced researchers had failed at previously; and who went his own way despite the skepticism of others. It would also be the story of a man whose extraordinary gallantry in World War I earned him a military honor and of a man who died too soon, tragically, in an airplane crash. It would be a powerful, romantic story.

Interestingly, that is the same impression a reader gets after reading his 300-page detailed biography, which describes day-by-day how insulin was discovered.

Frederick Grant Banting, or Fred, as everyone called him, was born on November 14, 1891 as the fifth child in the family of a farmer in the small town of Alliston. When he was younger, he was more interested in sports than studying. His school principal said that he was not someone they thought would become famous. His father wanted his son to be a minister, and as a good son, Fred acquiesced to his father's wishes and enrolled in college as a religion major. He came to realize after a year that it was not religion that made him happy, but medicine. With the help of the local minister, he was able to persuade his parents to let him study medicine at the University of Toronto, where he enrolled in 1912. Fred was an average student but a fervent researcher. He bought a microscope with his own money and studied everything he could, even a drop of blood from his own finger.

When World War I broke out, there was a shortage of surgeons on the front. Banting's class had their program condensed by two whole semesters, and in 1916, with no medical experience, he found himself in a military hospital in the English town of Buxton. It was not a bad duty. Behind the scenes, far from the front lines, he dressed the wounds and repaired the fractured bones of soldiers brought in straight from battle. But Fred did not like staying behind the scenes and thought he would be more useful closer to the front. He was transferred to the battle line in France.

During one of the final battles of the Great War, in September 1918, an artillery shell exploded not far from the first-aid station where Fred Banting was on duty, and he was seriously wounded in the arm by a piece of shrapnel. As he was the only doctor in the vicinity, he refused to be taken for treatment, instead bandaging his own arm and continuing his work. He worked nonstop for 17 hours. For his heroic actions, he was awarded the Military Cross.

The war ended and it was time to think about the future, but there were too many doctors in Toronto and finding a position was not easy. Banting decided to leave Toronto and establish a practice in a smaller town in western Ontario. He borrowed money from his father, bought a house from a former cobbler, hung out his shingle and waited. Patients were slow in coming and Banting started to rack up debt. He had plenty of time to study academic literature and was a frequent and recognized client of the local medical library, where they always set aside the latest issues of the medical magazines for him. This fact played a key role in his life and in the lives of all patients with diabetes. Banting's situation improved a bit after about three months, when the local university offered him a job as a part-time lecturer and demonstrator for $2 an hour. Not a fortune, but it was better than nothing.

Sunday, October 31, 1920 was the day when the incredible story-with-a-happy-ending of Frederick Grant Banting began. That evening, he was preparing his lecture on carbohydrate metabolism. It should be said that this was not one of his favorite subjects, nor was the associated disease, diabetes. For him, it was just the routine preparation of a lecture for the following day. When he finished, Banting went to bed with the brand-new issue of the magazine *Surgery, Gynecology and Obstetrics, Vol. 31, No. 5, 1920* that had been set aside for him at the library. The main article in the magazine was about diabetes and how it relates to the pancreas. The article itself was in no way special and offered no surprising findings or universal perspectives. The reason the article is important in the history of medicine is that Fred Banting read it that Sunday at 2:00 in the morning.

To understand the significance of this moment, we have to spend some time talking about the pancreas. We already know that it is a gland with two functions. One is the secretion of digestive juices containing enzymes into the digestive tract.

This function is called external secretion. The other function is the secretion of a hormone that was long suspected to be the culprit of diabetes. This has to do with internal secretion and attempts to isolate this hormone were unsuccessful for many years. The first problem was that the cells responsible for internal secretion comprise only a fraction of the total weight of the pancreas. The main hindrance was that in the process of isolation, these cells were destroyed by the digestive juices from external secretion. They knew the hormone existed and they knew where in the pancreas it was produced, but they could not isolate it.

The article Fred Banting read that Sunday night offered the solution. It described how to limit the production of digestive juices and thereby prevent the breakdown of the cells that produce the sought-after hormone. The pancreatic ducts would need to be surgically tied off, which would prevent the secretion of pancreatic juices. The laboratory animal would survive the procedure, but the cells that produce the juices would be destroyed. However, the cells that produce the unnamed hormone would remain unaffected. That night, Banting wrote in his notebook: "*Ligate pancreatic ducts of dog. Keep dogs alive till acini degenerate leaving islets. Try to isolate the internal secretion...*".

The very next morning he went to see his supervisor, who thought that Banting's idea was interesting, but not exactly groundbreaking. Besides, his department did not have the space or the capacity for experiments of this type. He would be better off looking elsewhere, like the University of Toronto.

A week later, on Monday, November 8, 1920, Fred Banting was sitting across from Professor John James Rickard Macleod in his office at the University of Toronto. Macleod did not know Banting and had only agreed to meet with him because he did not want to offend the young doctor. Banting, on the other hand, knew that Macleod was a respected scientist and experienced physiologist. On that day, neither one of them could have known that three years later, almost exactly to the day, they would become Nobel laureates.

Fred Banting had never been a persuasive speaker and that day may have actually been worse than normal. J. J. R. Macleod barely glanced at the papers on his desk. He later said of their meeting, "*I found that Dr Banting had only a superficial textbook knowledge of the work...and that he had very little practical familiarity with the methods by which such a problem could be investigated in the laboratory.*" The experienced professor told the young doctor that many accomplished scientists before him had tried to address the issue in well-equipped laboratories, and that it was a task that required the fullest, long-term dedication. The meeting did not end well. There was nothing left for Fred Banting to do but go back to his practice and wait for patients. Or he could try again. The next morning, he went to see Professor Macleod again.

We will probably never know what the senior scientist thought about that night, or what the junior scientist presented the next day, but what we do know is that this meeting was successful. Macleod promised Banting his help, including ten dogs, an assistant, and the use of a laboratory for eight weeks over the summer break. He must have thought it was worth a try. Even negative results would be of scientific value.

In May 1921, two of Macleod's students tossed a coin to see who would serve as Dr. Banting's assistant for the summer. Charles Herbert Best won (though some would argue he lost). A young, blond, blue-eyed, and handsome 22-year-old gentleman, he had been looking forward to spending the summer with his fiancée, Margaret. Considering how things worked out, Margaret must have forgiven him later.

Professor Macleod offered one more thing to Banting. Neither the young doctor nor his even younger assistant had any experience working with laboratory animals, so Macleod decided he would operate on the first dog himself. On May 17, 1921, he placed the first dog on the operating table, a brown bitch. He anesthetized her, opened her abdomen, removed a part of her pancreas, and closed her up again. This was intended to bring on a condition similar to diabetes. The entire operation took 80 minutes. After Macleod showed the young men how it was done, he went on vacation to Scotland.

The procedure planned by Frederick Grant Banting was relatively simple and needed at least two dogs. He would surgically tie off the pancreatic ducts of the first dog to kill off the digestive enzyme-producing cells, leaving only the cells that produce insulin. The process would take six to eight weeks. They would remove the second dog's pancreas to induce diabetes, and would then kill the first dog, remove the pancreas containing the insulin cells, and use the insulin to treat the dog with diabetes. If the second dog's blood sugar was lowered, the hypothesis would be verified and a treatment for diabetes would have been discovered.

When Frederick Allen, who wrote a 1,179-page monograph, was researching diabetes he needed 200 dogs and hundreds of other laboratory animals. When Banting and Best set off on their journey toward insulin, neither one had ever operated on a single animal, apart from what Professor Macleod had demonstrated for them. They were learning as they went. They certainly did not expect that every operation would turn out well, but the initial outcomes were nightmarish.

They started with a group of dogs in which they wanted to induce diabetes. The first dog died before the operation got underway of an overdose of anesthesia, while the second bled out during the operation. The third dog survived the operation but died two days later. In the meantime, the dog Macleod had operated on also died. The week came to an end with four dead animals. The fifth animal survived, and they finally had their first "diabetic" dog. They could now move on to the animals that would have the pancreatic ducts tied off, but once again, the first three dogs did not survive the operation. They found themselves in a situation where they had few of the ten dogs given to them by Macleod remaining, with no result in sight. They had to find an alternative way to obtain animals to continue with the experiments and so they started buying stray dogs for $3 each, using their own money.

But they were learning, and by early July, a month and a half after starting their research, they had seven dogs with ligated pancreases and two dogs with diabetes.

They were finally starting to feel that things would actually work out. They did not. When they opened up the dogs with the ligated pancreases, they discovered that in five of them the stitches had become loose and the pancreas had returned to normal. After repeating the operations, two of them died of infection, while the two that were to serve as the model for diabetes also died. When they tallied their losses, it was a disaster: out of 19 laboratory animals, 14 had died. They were left with only five dogs with ligated pancreases, and three of them needed repeat operations. The only excuse for this disaster is that the temperatures soared to 104 °F in the operating room that summer.

Banting and Best wiped the perspiration from their brows and kept at it. They needed an animal in which diabetes had been induced. A white, short-haired terrier mix survived the pancreatectomy, developed diabetes, and in late July was ready to undergo insulin therapy. The first ever.

They removed the degenerated pancreas from the first of five suitable dogs, carefully blended it and filtered off the solids, and injected the filtrate – a light pink liquid – into the white terrier mix. After two injections, the animal's blood sugar dropped to half of the initial "diabetic" number. Something in the filtrate worked.

Despite these advances at work, it was not a good time for Fred Banting. By leaving his practice, he had lost what little income he had. He earned a little money by performing a tonsillectomy on a friend and he sold surgical tools to raise meager funds. He let friends invite him for dinner and did not refuse meals offered by charitable organizations. In spite of everything, he and Charles Best worked day and night, repeating the well-established procedure on yet more dogs. When Fred Banting was down to his last seven cents, a professor at the University of Toronto took pity on him and arranged a part-time lecturer job for him with the princely pay of $250 per month. July passed by, then August, and soon it was September 21 – and Professor Macleod returned from his trip to Scotland. Despite the evident progress he was not overly impressed with the results, but he said the research should continue. However, Banting and Best needed space and money to continue their work, and although Macleod was not very accommodating at first, Banting was able to persuade him once again. Within a few days, the operating team had a new area that was much better equipped for their work. They also got a lab assistant to care for the animals. They could continue with their research.

But there was another problem, and it was related to the very essence of their research. The procedure for obtaining an adequate amount of the pancreas containing insulin was complicated and not always successful. It was also time-consuming, but most of all, if they wanted to save one animal (the one with induced diabetes) they had to sacrifice the other (the pancreas and insulin donor). That might have been justifiable in experimental medicine, but in clinical practice it was out of the question.

They needed to find another source of insulin-containing pancreases. One option was animal embryos, which do not have a developed digestive system but the pancreas would still be rich in insulin-producing cells. But where would they get animal embryos? Fred Banting was a farmer's son, and he recalled that farmers impregnated cows before sending them to slaughter because they would gain weight more rapidly.

The next day, November 17, 1921, exactly six months on from the first operation, Banting and Best went to visit the William Davies slaughterhouse in northwest Toronto. The employees could not help but wonder at the unusual request of these young men, but they had no objections and the scientists returned to their laboratory with eight calf embryos.

They now had adequate experience and adequate material, collected from the calf pancreases with a more effective method. They could proceed to the decisive step: administering insulin to a diabetic patient.

This is where the fourth member of the team comes in, biochemist James Bertram Collip. In spite of his young age – he was a year younger than Banting – he was already a skilled research scientist. With his ability to mix, filter, distill, vaporize, concentrate, dissolve, centrifuge, and re-mix, he later became something of a legend in Canadian medical circles. Banting and Best had occasionally run into him in the halls of the University of Toronto, and more than once the young biochemist had offered his help if they were ever to need it. They needed it now. The solution used to inject the laboratory animals was just a simple pancreatic extract containing a lot of impurities. It might have been fine for laboratory experiments, but it would need to be significantly purified if it was to be used on humans.

Everything seemed to be ready, and they could select the first patient. He was 14-year-old Leonard Thompson, who had been suffering from diabetes for two years and, as a result of Frederick Allen's therapy, weighed just 65 pounds at the time of his admission. His condition was more or less hopeless and he was completely lethargic. The first dose was administered on January 11, 1922 but the extract was impure and the effects were minimal, so much so that the doctors decided to discontinue therapy. It was time for James Bertram Collip to show what he knew. He worked day and night, and on the evening of January 22, he delivered a substantially more purified extract. The next day, January 23, the hopeless case Leonard Thompson received his first dose at 11:00 am and his second dose at 5:00 pm, with two more doses the following day. His blood sugar fell by three-quarters and the boy became more cheerful, active, and his condition was markedly improved. The first attempt to use the internal secretion of the pancreas in a human had been successful. Insulin was born.

It would still take a lot of time and a huge number of experiments before insulin could be administered on a large scale to diabetic patients, but that was just a continuation of the pioneering work that had been carried out by (in alphabetical order) Frederick Grant Banting, Charles Herbert Best, James Bertram Collip, and Professor John James Rickard Macleod. But who deserves the most credit? Which one, or more than one, is most deserving of the fame that came with this huge discovery? With the exception of Collip, every one of them declared in their memoirs that he himself was the most important, that he was the one who deserved most of the credit for the discovery of insulin. The Nobel Prize committee (the discovery alone deserved this recognition) had a difficult time deciding, but ultimately, they chose to give the prize to Banting and Macleod. Banting objected fiercely, believing that Best was more deserving of the prize than Macleod. He went so far as to give half of his prize money to his younger colleague. Not to be outdone, Macleod also shared his prize – with Collip.

It is interesting to see how the relationships developed between the characters in the insulin story. Of course, that is not what's important here. The important thing is that the young Leonard Thompson, the somewhat older Elizabeth Hughes, and the millions of patients who came after them, received the medicine that kept them from dying. They received the elixir of life.

Story 4.6: The strong-minded scientist and her four hands

For centuries, the natural sciences were the domain of men. While that dominance is no longer as considerable as it was in the early 20th century, men continue to be more predominant in the sciences. If we add up the number of Nobel laureates in chemistry from the time the prize was first awarded in 1901 until the last award in 2019, we get a total of 183 names. Only four of them are women: Marie Sklodowska Curie, her daughter Irène Joliot-Curie, Dorothy Crowfoot Hodgkin, and Ada E. Yonath. There is an even greater disparity between men and women laureates when it comes to physics. The Nobel Prize in physics has been awarded to 212 people; the only three women were Marie Sklodowska Curie, who received the award in both physics and chemistry, Maria Goeppert Mayer, and Donna Strickland. That is just six women out of 395 awards. These were strong women who succeeded in that male-dominated domain. This story is about one of them, Dorothy Hodgkin.

Biographers who write about the lives of scientists do not always find congruence between human traits and top-level science. Fortunately, this book offers several positive correlations. Charles Darwin often comes up as one of the greats of science and as a great human. He was extraordinarily friendly, a collegial coworker, supported young scientists, and was honorable and frank. He had no personal enemies. All of those epithets describe Dorothy Crowfoot Hodgkin perfectly. She was so kind and good it bordered on the naive. At the very least, her political persuasions lead us to think that.

She was born in Cairo on May 12, 1910, as Dorothy Mary Crowfoot. Her father, John Winter Crowfoot, was in the service of the English crown for the education ministry. He was tasked with overseeing the quality of education in Egypt, and later in Sudan. The entire family, including Dorothy's two younger sisters, lived in peace and relative luxury, first in Cairo and then in Khartoum, Sudan. John Crowfoot's job required frequent travel, and when he was away, Dorothy and her sisters were looked after by their mother and a nanny.

When the First World War broke out, the parents decided to send the children and the nanny to Europe where it was safer, while they remained in Khartoum. From that moment, Dorothy saw her parents only once or twice a year. Her father was appointed to a distinguished position in Khartoum and her mother accompanied him. Letters were the main method of communication between Dorothy and her parents throughout her childhood. She kept the correspondence and it is from there that the primary information about Dorothy's life can be drawn. It may appear that the behavior of the Hodgkins, as parents, was heartless, but at the time it was nothing out of the ordinary. There were many British colonies where royal subjects had to serve, and often these were places that were not very suitable for small children.

Dorothy grew up in a circle of family and her mother's friends, practically without a male role model. She grew up to be a young woman who was a bit bashful and indecisive but at the same time independent and ambitious. A timid, girlish smile graced a face with blue eyes and framed with golden hair. That smile is a typical mark of all her portraits, from the first ones where she was still a young woman, through to the final ones before her death.

Dorothy always wanted to be a chemist. At the age of 11, she was conducting experiments in the attic of her English home in Worthing, on the southern coast of the British Isles. She studied chemistry at the University of Oxford, graduating with a first-class honors degree as only the third woman ever to do so.

She studied for her doctorate at the rival Cambridge University. Although her sojourn there was relatively short – from the fall of 1932 to the fall of 1934 – it greatly influenced her life. The head of the laboratory was the charismatic John Desmond Bernal, who was ten years older than Dorothy, slightly hunchbacked, and with a wild mane of red hair. Ever since his youth he had been a staunch liberal, a member of the Communist Party of Great Britain, and was a great supporter and defender of the communist accomplishments in the Soviet Union until his death. This was nothing out of the ordinary, because at the time a large part of the English, and European, *intelligentsia* was liberal-minded. It comes as no surprise, with the menace of fascist and social nationalist ideologies sweeping across Europe, and civil war having broken out in Spain. In theory, the communist ideology was a positive alternative to the developments then playing out in western Europe. It would be a few decades before the practical realization of communism, with all its negative manifestations, would culminate. It was a time when King Oscar II of Sweden famously said: "*If you're not a socialist before you're twenty-five, you have no heart. If you are a socialist after twenty-five, you have no head.*"

Dorothy was not yet 25 years old and she did have a heart. Her heart was open not only to the ideologies of her boss, it was open to him personally as well. John Bernal was married, but that did not stop him from having a free-wheeling sex life, and that was something he did not keep to himself. He was involved in relationships with numerous women, of which Dorothy was one.

John Desmond Bernal influenced the life of Dorothy Mary Crowfoot politically and emotionally, but that was not all. He was an expert in a field known as X-ray crystallography. Dorothy decided to pursue this discipline at Cambridge and later became known for her expertise in the field. But first it is important to explain what crystallography is and how X-rays are a part of it.

A simplified definition is that crystallography is an experimental science that studies the arrangement of atoms in solids. This enables the determination of the atomic and molecular structure of crystals. X-ray crystallography is a method that uses X-rays to study the atomic and molecular structure of crystals. As the name suggests, the substance that is studied must be available in crystal form. This is not a big problem with inorganic and small organic compounds; we encounter crystals in table salt and glucose in granulated sugar every day. We can also stare in wonder at the cluster of ice crystals in snowflakes until they melt in our palm. The problem arises when the molecule is bigger, which is why obtaining suitable crystals plays a key role in X-ray crystallography.

The crystal to be studied – which must be at least one-tenth of a millimeter in size – will then fit into the X-ray beam. As the beam passes through the crystal, it diffracts in different directions. By measuring the angles and intensity of the diffracted beams, the positions of the atoms can be determined and thereby so can the structure of the substance. This requires precision work as well as a great deal of calculations. Modern computer technology makes this final step much easier than it was in the early 20th century, when all the calculations were done by hand.

Dorothy joined John Bernal's team at just the right time. They had just begun to study the structure of biological molecules in one of the first ever crystallography laboratories. They began with a group of sterols that contained dozens of atoms in their molecules; cholesterol belongs to this group. The first huge success was when they obtained good quality X-ray photographs of a true biological macromolecule: the enzyme pepsin. Pepsin is located in the stomach and helps to digest proteins.

After obtaining her doctorate degree, Dorothy returned to her "native" Oxford. Despite her young age – she was barely 25 at the time – she became a renowned expert in X-ray crystallography. She was so well-known that she came to be in possession of a crystalline substance that had been discovered not ten years earlier by the young Canadians Fred Banting and Charles Best; crystalline insulin. This is how the story of Dorothy, then still Crowfoot, became a part of this chapter, when she viewed the minute insulin crystals under a microscope for the first time on October 25, 1934. She could not have known at the time that it would be another 35 years before she would be able to show the world their structure.

Life goes on. In the spring of 1937, she met Thomas Lionel Hodgkin, the descendant of a family of educators with a long tradition in medicine, who was an idealistic romantic and member of the Communist Party. Thomas later became a recognized expert on Marxism and African history. On December 16 of the same year, Dorothy Crowfoot became Dorothy Hodgkin, but she kept her maiden name to use on her scientific papers and was still addressed as Miss Crowfoot, even at an advanced stage of pregnancy. In 1938, she gave birth to the first of their three children and became the first woman at Oxford to be given paid maternity leave.

Dorothy continued her research, studying a wide variety of macromolecules, but she kept coming back to insulin. Two important things happened to her in the fall of 1941. Her daughter was born in September, and in November she was given an important task. She was invited, as an experienced scientist, to participate in a research project that was seeking to solve the structure of the "wonder drug," penicillin. They were successful. Thanks to this, Dorothy Hodgkin might actually have found herself not in one, but in two chapters of this book – this chapter about insulin and the following chapter about penicillin.

The successful identification of the structure of penicillin secured her reputation as an internationally recognized scientist, and in 1948 she was given another task: identify the structure of vitamin B_{12}. It took an entire seven years, but she once again succeeded. By May 1955, she had verified practically the entire structure of this vitamin.

Virtually none of the experts in crystallography in the 1950s would have expressed any doubt that Dorothy Hodgkin was the "first lady" of this field. Her long years of successful work, and her achievements in identifying the structures of penicillin and vitamin B_{12}, earned her numerous awards, including fellowship in the Royal Society. It was generally expected that she would win the Nobel Prize for discovering the structure of vitamin B_{12} but she did not win it in 1956, or in later years. She was nominated for the prize for both chemistry and physics, because X-ray crystallography lies somewhere between these two sciences. In 1964, when she had all but given up hope, and nearly ten years after her greatest discovery, Dorothy Hodgkin finally won the Nobel Prize for chemistry. The reason for the long wait is probably best demonstrated by the newspaper headlines announcing that she had won the prize: "Nobel Prize for mother of three," and "Norfolk girl wins Nobel." They described her as "*an affable-looking housewife*" who won the prize "*for a thoroughly unhousewifely skill...*" In the view of society in those days, her skills were so extraordinary that they gave preference to the typical scientist – a man.

Dorothy still had many more questions about the structure of insulin. Compared with other molecules whose structure had already been described, insulin was much smaller. It should not have been difficult to unravel its structure, but insulin comprises more than one smaller structure linked together in a complex three-dimensional symmetry and it persisted in remaining a mystery. Not even newer and more efficient instruments and major improvements in computer technology offered any help, nor did the rising number of scientists in various laboratories dealing with the structure of insulin. Another five years would pass from the time she won the Nobel Prize, but in 1969 her team was finally able to announce that they had determined the complete three-dimensional structure of insulin.

In the late 1960s, Dorothy Hodgkin gradually brought her active research to an end and began to travel. She travelled practically the entire world, lecturing on crystallography and on war and peace. She had been strongly liberal-minded since youth, her first tutor was a communist and so was her husband. Although she never formally became a member of the Communist Party, her political activity was strongly tied to the ideology. She condemned social inequalities and wanted to prevent armed conflict through negotiation. She chaired numerous international peace conferences and strived for peace talks between the East and the West. The Soviet Union honored her efforts by awarding her the Lenin Peace Prize, a sort of Soviet counterpart to the Nobel Prize. Perversely, the United States would not give her an entry visa. However, her conceptions of communism and socialism were more idealistic than ideological and at times bordered on the naive. That is the only way to explain why, at a time when she was still active in her career, she participated in the celebrations of the tenth anniversary of the revolution in China. It was as though she was unaware that the celebrations were being held just two years after Mao Zedong initiated the Cultural Revolution which, among other things, caused the stagnation of scientific development in China for many years. She sat together with the highest officials of the then-socialist camp, including Nikita Khrushchev, Zhou Enlai, Ho Chi Minh, and Mao himself. She even wrote the foreword to a scientific work written by Elena Ceausescu, the wife of Romanian dictator Nicolae Ceausescu, praising the author's excellent achievements and impressive career. It never occurred to her in her good-hearted naivety that the book's author had never even finished high school and that the book was a complete sham, written in the name of comrade Elena by an entire team of scientists.

Dorothy Mary Crowfoot Hodgkin was a truly remarkable person. Throughout her career, she had the ability to address problems with a laser focus and she did not back off until she had found the solution. She was patient and always optimistic. She knew how to communicate with people and that skill helped her form a huge circle of close friends and distant co-workers. While many of her political activities can be criticized, her good faith and sincerity cannot.

The most famous portrait of Dorothy Hodgkin hangs in the National Portrait Gallery in England. It depicts an older, gray-haired woman sitting behind a desk, wearing black-framed glasses and buried in her work. On the wall behind her is a shelf filled with scientific publications, while the large desk behind which she sits is covered with scientific journals and three-dimensional models of insulin and vitamin B_{12} molecules. Dorothy is depicted as having four hands. In one, she holds a magnifying glass, in another she is looking at some paper or other, and with the other two she is creating some complex diagram on the desk. Two messages are contained in this unusual portrayal. The first conveys Dorothy Hodgkin's industriousness but the second only appears after examining the portrait more closely. Her hands are deformed by rheumatoid arthritis and the depiction of the four hands in various positions emphasizes the deformation.

Dorothy began experiencing pain in her upper extremities when she was just 20 years old, and she was diagnosed with rheumatoid arthritis. This is an inflammatory disease that mainly afflicts the joints. It is accompanied by severe pain and arthritic changes in the joints, often resulting in complete disability – as was the fate of Dorothy. She was forced to live out her last years in a wheelchair, but not even that prevented her from travelling, lecturing, organizing, and spreading her optimism. She died on July 29, 1994.

This story ends with a quote from Dorothy Mary Crowfoot Hodgkin herself: "*There are two moments that are important. There's the moment when you know you can find out the answer and that's the period you are sleepless before you know what it is. When you've got it and know what it is, then you can rest easy.*"

Concluding remarks

Most of us are fortunate enough to have our own "internal" insulin. Those who do not need external insulin. Insulin is therefore not a drug that cures diabetes; a diabetic remains a diabetic for life. The administration of "external" insulin only provides the body with a hormone that, for reasons as yet unclear, the body does not produce on its own, and without which life is not possible. Insulin has saved the lives of millions of people around the world.

We have a medicine that can save the life of a person with diabetes, just like it saved the life of Elizabeth Hughes. This makes it so much more incredible that more people die today of diabetes and the complications associated with it than did at the time when Frederick Grant Banting, Charles Herbert Best, James Bertram Collip, and John James Rickard Macleod discovered insulin. Severalfold more people. According to data from the International Diabetes Federation, diabetes causes approximately five million deaths per year. How is that possible?

Diabetes was already known to the ancient Egyptians as a relatively rare disease that manifests with frequent urination and weight loss. In the 17th century, they knew it as a serious but rare disease. In 1922, at the time of the discovery made by those scientists, diabetes was still a relatively rare disease. Roughly two million diabetics, maybe fewer, needed insulin at the time. In 2016, less than 100 years later, there were 422,000,000 worldwide cases of diabetes. Four hundred and twenty-two million. The most significant increase in diabetes has occurred in the last few decades, with the number of people with diabetes having quadrupled since 1980. In developed countries, about one in ten has diabetes; in the U.S., it is one in eight. However, the rising prevalence of the disease in developing countries would seem to indicate that by 2030, the majority of diabetics will be living in African and Asian countries. A rise in the standard of living goes hand in hand with the adoption of the bad habits we have in the developed world.

These numbers show that we are in the midst of what you could call a diabetes epidemic. Unlike other epidemics, where the main problem is finding a medicine to treat the disease, we *already have* that medicine, and we have had it for nearly 100 years. Yet the prevalence of this disease has risen by over two hundredfold.

It is wonderful that we have a medicine to treat diabetes, thanks to which the majority of sufferers are able to live, but neither insulin, nor other anti-diabetics, will stop the rising prevalence of the disease. The vast majority of people have type two diabetes. It is not caused by bacteria or a virus, or any other outside factor. In most cases, we only have ourselves to blame.

5

Penicillin

The legend of how penicillin was discovered is downright mythological. It goes something like this: One day, British scientist Alexander Fleming returned from a long vacation and found, among the clutter on a table in his laboratory, several glass plates overgrown with colonies of bacteria. But one of the plates was different. He soon determined that the plate was contaminated with a mold that had killed off the bacteria. He named the mold *Penicillium*, isolated penicillin from the mold, and that is how the first antibacterial therapy was born.

Only one part of that legend is true: that Alexander Fleming returned from vacation.

Story 5.1: The doctor with stained hands and the magic bullet

You would be hard put to find many scientists in all of history who better epitomized the stereotype of this profession than Paul Ehrlich.

His work was his entire life. He was oblivious to his surroundings, he was absent-minded, forgetful and restless. He wore horn-rimmed glasses and smoked 15 cigars a day. To make sure he had one handy, he always carried a supply under his arm. He said of himself that he had blinders on his eyes so that he could concentrate exclusively on one thing; he called himself a *monomaniac*. Ehrlich carried a chemistry encyclopedia around in his head, but had little knowledge of other areas. He was not interested in music, literature, or any other serious form of art, although he loved the stories of Sherlock Holmes, maybe because the two men had something in common. The protagonist of Arthur Conan Doyle's stories also

V. Marko, *From Aspirin to Viagra*, Springer Praxis Books,
https://doi.org/10.1007/978-3-030-44286-6_5

believed that his brain was like an attic that he did not intend to clutter up with useless things.

By contrast, Professor Ehrlich's study was entirely cluttered. Books and magazines were strewn everywhere, on the settee, on chairs. Stacks of academic literature covered every inch of the rug. Spots that were not covered in literature contained test tubes, vessels, and other various and sundry laboratory glass.

He was obsessed with discussing his work with anyone who was willing to listen, and his monologues could be hours long. If he needed to explain one of his many hypotheses, he would not hesitate to draw a picture of it on his co-debater's cuff. He did not think twice of drawing on doors, or getting on his knees to scribble diagrams on the floor of his laboratory. At home, he drew on tablecloths, much to the chagrin of Mrs. Ehrlich. Fortunately for both of them, her father, the industrialist Joseph Pinkus, owned a large textile factory and so the family always had a supply of clean, new tablecloths.

Ehrlich's forgetfulness is the subject of various humorous stories. He would send himself postcards to remind him of the birthdays and anniversaries of people close to him, and if he happened to take documents to work with him in an envelope, he would write his address on the envelope with a plea to return it for a reward. That paid off more than once for him.

Professor Paul Ehrlich was truly a very interesting and unusual character. Probably the most likeable thing about him was that, despite all that he achieved, he was not at all concerned with his own dignity.

He was also a person with an amazing ability to sink his teeth into any problem he was looking to solve, an ability to pursue his goals. He went down many dead ends with unyielding energy until he made the discovery that made him famous: compound 606, the "magic bullet," Salvarsan.

Paul Ehrlich was born on March 14, 1854, in Strehlen in Lower Silesia (now known as Strzelin in Poland's Lower Silesian Voivodeship), to a middle-class Jewish family. His father and grandfather were popular local innkeepers and liquor distillers. He was not top of his class in either elementary or high school, and he barely passed his German finals. He studied medicine at a few different universities, in Breslau, Strasbourg, and Freiburg, and obtained his doctorate at the university in Leipzig.

Ehrlich was not an exemplary student, skipping lectures and leaving textbooks unread. Some of his professors had doubts as to whether he would even graduate from medical school. At that time, he was already deeply involved in an unusual activity, one that would become his destiny and accompany him throughout his entire life: he discovered the possibilities of staining biological specimens. Most biological specimens, whether cells or tissue, look relatively featureless under a microscope. Because only various shades of gray are visible, the individual parts of the specimens are difficult to discern. As a student, Paul Ehrlich discovered that the various parts of cells and tissue absorb dyes to varying degrees, making them colorful and more visible under the microscope and thus easier to discern. He began to experiment, gathering all the available dyes: purple tincture of iodine; methyl violet; red-purple fuchsine; red-yellow purpurin; red safranine; and many others. He prepared hundreds of specimens and stained his heart out. He could easily be told apart from the other students, because his arms were always stained up to his elbows.

Ehrlich managed to graduate from university. He wrote a doctoral dissertation entitled "Contributions to the Theory and Practice of Histological Staining" (*Beiträge zur Theorie und Praxis der histologischen Färbung*), and became a highly regarded authority in specimen staining.

His interest in staining cells did not wane when he finished school and he developed staining techniques that allowed him to make visible previously unknown types of blood cells. He also differentiated the known types in more detail, including red and white blood cells. These were not discoveries for the sake of discovering. His work resulted in a better capability of studying the immune system and diagnosing anemia and leukemia. He also discovered a simple test that could distinguish typhoid from simple cases of diarrhea, which contributed to the diagnosis and treatment of a serious infectious disease. That is not even close to the last of his discoveries, and his stained arms continued to be his identification mark.

The turning point in Ehrlich's life came in 1891, when the renowned bacteriologist Robert Koch invited him to come and work at his Institute of Infectious Diseases (*Institut für Infektionskrankheiten*). This began the second phase of the professional career of Professor Paul Ehrlich, during which he became an immunologist for 15 years. He was no ordinary immunologist.

The subject of his first research was an antiserum for treating diphtheria and tetanus, one of which, the antiserum for diphtheria, found its way from the laboratory into medical practice in 1894. The research was so groundbreaking that his coworker, Emil von Behring, received the Nobel Prize for it in 1901. Paul Ehrlich felt overlooked at the time, as he felt that he had also deserved the prize, but his time was yet to come.

Although he still had to wait a while for the Nobel Prize, his career as an esteemed physician and researcher skyrocketed. In 1897, he was appointed as Senior Privy Medical Councilor (*Geheimer Obermedizinalrat*) to the court of the Prussian King Wilhelm II. This was a very high position that earned him the right to be addressed as "Excellency." When the Institute for Serum Research and Testing (*Institut für Serumforschung und Serumprüfung*) was established in Berlin, it was inevitable that Professor Paul Ehrlich would become its director.

The institute was later renamed the Royal Prussian Institute of Experimental Therapy (*Königlich Preußisches Institut für Experimentelle Therapie*) and relocated to Frankfurt, and the director moved with it. It was in Frankfurt that Ehrlich made his greatest discoveries. He formulated the principles of immunization and laid the theoretical foundations of immunology that are still recognized today. The residents of Frankfurt were rightfully proud of "our Professor Ehrlich." Due in part to his eccentricity, he became a well-known and well-loved man there. When he was finally awarded the Nobel Prize in 1908, the city council organized a spectacular Ehrlich Day event, attended by science heavyweights from many of Germany's universities and by the highest representatives of the city.

The third phase of his career began when the widow of Frankfurt's banking wizard, Georg Speyer, endowed a new science institute in Frankfurt in honor of her husband. His Excellency Professor Paul Ehrlich was asked to be its director. The new institute was built next door to the Royal Prussian Institute, and as the director of both, Paul Ehrlich was able to go from one workplace to another with ease. While he continued in his previous research at the old institute, in the new one he had the opportunity to return to his passion for "staining," out of which came the discovery of a new drug.

It happened during an experiment, when Ehrlich used a living organism instead of a dead specimen. He injected a small amount of his favorite dye – methylene blue – into the ear vein of a rabbit and was surprised that the dye worked selectively, staining the nerve endings blue but not affecting the remaining tissue. It was just a small step from this discovery to the idea that if there are dyes with a special affinity with only certain tissues, organs or microorganisms, then there must be other substances like dyes with a similar affinity. If a substance could be found that, in addition to having an affinity, also acted as a sort of poison for a harmful microorganism, then the microorganism could be targeted for destruction. Now all that was left was to find that substance. He decided to give it a generic name – the "magic bullet" *(die magische Kugel)* – and began his search. Eight years would pass before he could announce that "*die magische Kugel*" actually existed. It was named Salvarsan and it was used to treat one of the deadliest diseases, syphilis.

However, before he found his way to the microorganism that causes syphilis, Ehrlich began with a different aim for his magic bullet: the microorganisms that cause sleeping sickness, called trypanosomes. They were easy to stain and were large enough to view under the microscope without difficulty. He had no idea how to proceed, so he decided to go with the oldest known method of problem-solving, trial and error. He ordered hundreds of mice and began experimenting with different dyes – 500 of them, to be precise. He was partially successful with some, but there were no definitive results. All the experiments ended in one huge disappointment. Professor Ehrlich's stains were ineffective.

At the beginning of this story, we mentioned that one of Professor Ehrlich's strongest characteristics was his inexhaustible stamina. When one group of compounds did not work, he decided to try again with arsenic-containing molecules. Once again, he ordered hundreds of mice and began a new series of experiments.

His coworkers prepared hundreds of new compounds and tested each one in various concentrations on dozens of mice. They did thousands of experiments.

In the meantime, the microorganism that causes syphilis, the terror of the early 20th century, was discovered in 1905. One-sixth of all Parisians and over ten percent of Londoners were infected with syphilis. Because the bacteria that cause syphilis – spirochetes – were initially thought to be similar to trypanosomes, Paul Ehrlich added this microorganism to his testing. Again, he ordered hundreds more mice and his coworkers conducted thousands more observations. Every new compound that was tested in experiments with trypanosomes and spirochetes was given a unique number. The first 605 were ineffective, but the counting ended on August 31, 1909, with number 606.

That is how the therapy that would go down in history as Compound 606, or Ehrlich's magic bullet, was invented. It was later given the name Salvarsan and became the first chemotherapeutic, the first substance specifically targeted to treat diseases caused by microorganisms. It helped thousands of syphilis patients. Syphilis was no longer a nightmare, no longer a punishment of sorts for those who engaged in sexual intercourse.

Salvarsan turned His Excellency Professor Paul Ehrlich into a real celebrity. The people of Frankfurt knew and were proud of him, but with his typical humility he said of his discovery, "*for seven years of misfortune I had one moment of good luck.*"

Decades later, Salvarsan was replaced by the more effective and safer penicillin and it is no longer used. It was, however, the first laboratory-prepared drug for the treatment of bacterial diseases, which is why it deserves a place in the history of medicine – as does Paul Ehrlich.

Story 5.2: The rejected Nobel Prize and saving young Hildegard

In 1913, four years after the discovery of Salvarsan, Professor Paul Ehrlich stood before the crowd at the 17th International Congress of Medicine in London. He delivered a keynote address on the subject of antimicrobial therapy, in which he predicted that the collective efforts of scientists over the next five years would bring groundbreaking advances in the treatment of infectious diseases. He believed there was already sufficient knowledge of the diseases. He said in his speech that "*…this great International Congress, to which thousands have been drawn from all lands, [would] bear testimony to the fact that in the world of science all national barriers have fallen.*"

These were optimistic words; too optimistic, as we now know. Less than a year later, the nations of Europe entered into a long and horrifying war that wiped out any efforts to eliminate barriers, and not just barriers in science.

It would be a full 20 years before Ehrlich's predicted breakthrough actually occurred. The wait finally ended on Christmas Day in 1932, when the huge dye industry syndicate known as IG Farben filed a patent application for a crystalline red powder with the chemical name *4-sulfonamide-2,4-diaminobenzene*. This substance was later named Prontosil and it became the first of a class of drugs known commonly as sulfonamides. These were revolutionary in the treatment of infectious diseases. They did not just help save the lives of millions of people, they also gave doctors a weapon to combat a multitude of deadly diseases. They brought an optimism and faith in medical miracles never before seen.

There were many reasons why the search for these additional "magic bullets" took so much longer than Paul Ehrlich had predicted in his speech. The main reason was that there was no cohesive theory to serve as the basis for finding effective substances. Even Ehrlich's approach was purely empirical. He believed that some dyes would be effective against harmful microorganisms, and then he just started searching and experimenting. He was the "Last of the Mohicans", as it were; one of the last great scientists whose authority was so great he was able to persuade others that the given aim was the right aim.

But the future already belonged to a different approach. There were now pharmaceutical companies with teams of chemists, pharmacists, biologists, doctors and scientists of numerous other specializations. The new substance was the achievement of the collective efforts of many people, each of whom contributed in their own way to the resulting product, so it was often difficult to determine who deserved the most credit. New drug research was slowly becoming institutionalized and industrialized.

One of the first products to come out of this new approach was Prontosil. As noted earlier, this drug was invented by the giant IG Farben syndicate. The full name of the chemical and pharmaceutical conglomerate was *Interessen-Gemeinshaft Farbenindustrie* and it was formed in 1925 by the merger of several chemical companies, the largest of which were BASF, Bayer, Hoechst, and AGFA. It was the largest chemical concern of its day and the fourth largest company in the world.

Prontosil was the first success recorded by the research division of the pharmaceutical company, and although it was the achievement of an entire team of researchers, the bulk of the credit for its development was given to German bacteriologist Gerhard Domagk. He was awarded the Nobel Prize in 1939 for the discovery of Prontosil.

Gerhard Johannes Paul Domagk was born on October 30, 1895, in an eastern Prussian village called Lagów that is now a part of western Poland. He went to medical school, but his studies were interrupted by World War I. He enlisted, was wounded, and was then transferred to serve as a medic in the cholera barracks in Russia. The time he spent in Russia seems to have been the reason he chose his later medical specialization. He saw how powerless the doctors were when faced with patients with cholera, typhoid, and other infectious diseases. There was no cure for any of those diseases at that time.

Domagk joined IG Farben in 1927, and by 1929 he was the director of the newly established Laboratory of Experimental Pathology and Bacteriology, which tested the antibacterial properties of various dyes following the example of Paul Ehrlich. Compared to Ehrlich, however, they had the huge advantage of being able to tap into the knowledge of the chemists in the company. As the name suggests, IG Farben was a company that developed and manufactured dyes for industrial use. Just like their predecessor, Gerhard Domagk's team tried several hundred various dye compounds before they hit on the right one. According to Domagk's own laboratory logbook, it happened on December 20, 1932, the day he injected compound number KL695 into mice that had been infected with deadly bacteria. The laboratory animals survived.

As soon as the patent application was filed, extensive laboratory testing began to verify the results of the original experiment. The results were good, and in early 1933 they were able to move on to clinical trials. The drug was administered to patients suffering from various bacterial infections and it was effective almost immediately in the majority of the patients. The first publications describing the "magic" properties of the new drug appeared in 1935. One of the first reports was the somewhat personal case study of Gerhard Domagk. It concerned his six-year-old daughter, Hildegard, who had accidently stuck herself with a needle at Christmas. She developed a bacterial infection and became severely ill with a high fever, and the infection gradually moved up through her entire arm. In those days, the only way to resolve similar infections was by amputation, but in her case even an amputation would not have guaranteed her survival. Calvin Coolidge, Jr., the son of American president Calvin Coolidge, had faced a similar problem ten years before Hildegard was injured, when an infection spread through his body through a blister on his toe. He died within days.

Hildegard was one of the lucky ones who lived on the cusp of two periods of time; one in which there was nothing available to help her, and the other, better period of time where a solution was found. A magic solution. All it took was for her father, Gerhard, to administer a new therapy, even though it had not yet been fully tested. Just like the laboratory mice a year or so earlier, young Hildegard survived the infection without severe consequences.

As soon as the first reports were published about the beneficial properties of Prontosil for the treatment of bacterial infections, it became a huge success for IG Farben. In 1935, in the first year after it was authorized for the market, it made the company 175,000 Deutsche marks. In 1936, it was up to a million, and in 1937 it grew to five million. By 1935, it was also available in France, a year later in Great Britain, and the year after that it became available in the U.S. The success of the new drug was not dampened by the fact that Prontosil was originally a red dye and that after it was administered to patients, they would turn red. Fortunately, the redness faded once treatment ended.

Prontosil was the first drug to treat various types of bacterial infections in the human body effectively. Unsurprisingly, in 1939, Gerhard Domagk received the Nobel Prize in physiology and medicine for this groundbreaking discovery, although this prize was surrounded by much controversy. At the time, German citizens were forbidden to accept the Nobel Prize by a decree written by Adolf Hitler himself, in response to the Nobel Prize in 1935 being awarded to German pacifist (and enemy of the Third Reich), Carl von Ossietzky. Three weeks after the announcement that he had been awarded the Nobel Prize, the Gestapo surrounded the home of Gerhard Domagk. They confiscated all documents related to the prize and hauled the scientist off to prison where he spent an entire week, during which he was forced to explain numerous times the circumstances of how he came to be awarded the Nobel Prize.

Even though the time Gerhard Domagk spent in prison was unpleasant, there is one humorous anecdote associated with it. When one of the prison guards asked why he was in prison, Domagk replied: "*Because I was awarded the Nobel Prize.*" A little while later, he heard the guard telling a colleague that the man in his cell was crazy.

When he was released from prison, Gerhard Domagk's troubles were far from over. The Gestapo detained him again, and as a sign of loyalty to the Third Reich he was forced to sign a pre-written letter strictly rejecting the Nobel Prize. He would collect his prize in 1947, after the war had ended.

The discovery of Prontosil brought with it the institutionalization of scientific research, but it did not diminish the importance of serendipity, which often led the institutionalized research parties to new and groundbreaking discoveries. One such serendipitous event took place on November 6, 1935, in the research laboratories of the Pasteur Institute in Paris. That morning, ten groups of laboratory mice, into which a deadly bacterial infection had been introduced, were prepared for an experiment. One was the control group, which would not receive any drug, while the remaining nine groups would serve to test nine potential new substances. However, the researchers only had eight new molecules. In order to use all the laboratory animals, they added a ninth test substance, an intermediate product occurring in the preparation of Prontosil and the substance from which Prontosil arose. To their surprise, the next day they discovered that the most active of all the tested substances was that particular intermediate product. It was later discovered this product, named sulfanilamide, was actually the active substance even in the administration of Prontosil and similar molecules. They break down in the body into sulfanilamide, which then effectively kills the bacteria. We saw a similar mechanism with aspirin. The tablet you take is aspirin – acetylsalicylic acid – but in your body it turns into salicylic acid and that is actually what causes the drug to take action.

Sulfanilamide continued in the groundbreaking treatment of bacterial infections that was started by Prontosil and it had several advantages over its predecessor. It had a simpler structure, it was colorless (patients didn't turn red after using it), and…it was unpatentable. It was a well-known compound first synthesized in 1908 and any halfway competent pharmaceutical manufacturer could apply it to the treatment of bacterial diseases. Many more similar molecules based upon it were created, and by the end of the 1930s hundreds of manufacturers were producing thousands of tons of sulfonamides.

The fact that practically any capable chemist could produce sulfonamides also led to negative consequences. The biggest tragedy took place in 1937 and was caused by a pharmaceutical company called the S. E. Massengill Company of Bristol, Tennessee. In their efforts to prepare a liquid form of sulfanilamide, they dissolved the substance in a mixture of water and diethylene glycol. They called it "Elixir of Sulfanilamide," and the syrup quickly became popular, but the consequences were tragic. Over 100 people died of diethylene glycol poisoning, a large number of which were children. The final victim of the entire incident was the actual chemist who had prepared the Elixir. He was so consumed with guilt that he killed himself.

The discovery of sulfonamides represented one of the greatest advances in the history of medicine. For the first time, doctors were not defenseless against infectious diseases. Thanks to sulfonamides, the incidence of infectious diseases began to decrease and so did the mortality rate of infections. In the first 15 years after the discovery, from 1937 to 1952, they helped save over 1.5 million human lives. The incidence of pneumonia, flu, typhoid, dysentery, and many other infectious diseases fell. Minor, and even more serious injuries no longer carried risk of death because of an infection. While most patients were hospitalized for infectious diseases in the 1930s, by the early 1950s such patients had been replaced by those with non-infectious diseases – what are now often called civilization diseases. Penicillin-based antibiotics began forcing out sulfonamides in the 1940s, but they remain a part of modern medicine.

The life of British Prime Minister Winston Churchill was also saved by sulfonamides. In 1943, he traveled to Tehran to meet with Stalin and came down with a bad case of pneumonia. Before the discovery of sulfonamides, and considering the age and exhaustion of the patient, it would have been fatal. But due to the new drug, he soon recovered from the pneumonia. Sulfonamides may therefore have actually influenced the course of World War II.

Their contribution is not just limited to being effective. The discovery of sulfonamides changed medicine as a whole. As with the case of little Hildegard, doctors and medical professionals did not just have to stand by powerlessly and watch patients suffer – and frequently die a slow death. Sulfonamides turned hopelessness into hope.

Story 5.3: The Scottish bacteriologist and his return from vacation

Humans have always believed in miracles. They were and are a part of religion and worship. In the past, they were the only way to explain many of the events people encountered in their lives. They looked to miracles in times of poverty and sickness, and before the causes of diseases were discovered, miracles were the only way to deal with them.

Gradually, over time, "non-miracle" treatments were added to the miracles. In the mid-17th century, the citizens of Rome and visitors to the eternal city received help in the treatment of malaria from the quinine contained in Jesuit's bark. In the Battle of Trafalgar, Admiral Nelson was able to rely on vitamin C from citrus juice. If anyone needed to reduce a fever that came with the flu, starting in the early 20th century they could reach for an aspirin. But a group of diseases remained against which there was no effective treatment, diseases that since the dawn of humankind had been most often responsible for human death. Infectious diseases.

It is more than chilling just to name the biggest epidemics in human history that were caused by infectious diseases. The Plague of Justinian (541–750 AD) decimated 50–60 percent of the population. From 1347 to 1352, the Black Death, also caused by a bacterial infection, slashed Europe's population from 450 million to 350 million. The smallpox and typhoid fever brought by European explorers to Central and South America caused Mexico's population to drop from 20 million to three million between 1518 and 1568. Smallpox killed 400,000 Europeans a year in the 18th century, and a quarter of the adult population died of tuberculosis in the 19th century. For generations, scarlet fever, meningitis, pneumonia, rheumatic fever, cholera, and tuberculosis were considered a death sentence.

It was not only during epidemics that people suffered and died due to infections. Venereal diseases and pneumonia, puerperal fever, typhoid fever, diarrheal diseases, gangrene resulting from injury or surgery – all of these were frequent visitors in practically every family. There were no miracles that could help to fight these and the many other infectious diseases. But Paul Ehrlich's Salvarsan and Gerhard Domagk's Prontosil indicated that a miracle could become a reality. And it did.

It all began on Monday, September 3, 1928, on a table in the laboratory of one Scottish bacteriologist who had just returned from a month-long vacation. His is one of the most famous names in the world of medicine and natural sciences: Alexander Fleming.

His eventful and serendipitous life began on August 6, 1881 on a farm not far from the town of Darvel in the west of Scotland. He was the seventh of eight children. Alec, as he was known at home, was only seven years old when his father died and he was raised by his mother and older brothers. When he was 14, he went to live with one of his brothers in London and attended a business academy. The gifted and studious Alec finished school in two years, and at the age of 16 he took a job with a shipping company that arranged transportation by steamer between Europe and America. He did not enjoy work as a clerk one bit but fortune smiled on him (not for the last time) in 1901, when he inherited a small amount of money from an uncle. He decided to invest the money into the study of medicine and that same year he enrolled in St. Mary's Hospital Medical School. Around this time the story of Alec ends, and the life of Sandy, as Alexander Fleming was known to his friends, begins.

Sandy was a Scot who was not very talkative, but all the more stubborn. He was rather short, with blond hair and blue eyes. He was left with a broken nose after a childhood injury and it made him look a bit like a bantam-weight boxer, but he had a gentleness, calmness, and unobtrusiveness about him. His only quirk was the colorful bow ties he wore instead of regular ties.

He took to the life of a student right away. Fleming was a gifted scholar and did not have to spend much time with his nose buried in textbooks, but even so he won one award after another and there were very few subjects in which he did not achieve distinction. The less time he spent reading textbooks, the more he was able to spend on his favorite sport. He became a keen member of the rifle club at St. Mary's Hospital Medical School and helped the club to win the Hospital Cup. However unlikely it may seem, it was his marksmanship – and one other serendipitous moment in his life – that made him become a bacteriologist. When Fleming graduated, it looked like the rifle club would lose its best member. By coincidence, a spot opened up for a bacteriologist in the inoculation department at St. Mary's Hospital. The captain of the rifle club heard about it and suggested that Fleming apply for the position. Even though he had not intended to become a bacteriologist, he accepted the job. The rifle club kept its great team member and humanity got penicillin. This was in 1906, still more than 20 years away from that memorable Monday in 1928.

In the meantime, Sandy became Flem. This was the name by which his coworkers in the department knew him. He gradually settled in to bacteriology and earnestly worked on the development of vaccines. He was quiet and unassuming and was not an impassioned speaker, even among great debaters. His lectures were a "boring" itemization of outcomes and his contributions to discussions were rare. His complete inability to engage the students meant that a large part of his pioneering achievements and bold theories went largely unnoticed by his contemporaries.

During World War I, Fleming and his colleagues from the inoculation department were transferred from London to northern France to study the infections that caused soldiers on the front lines to become sick and die. They were confronted with the fact that almost ten million soldiers – more than half of those who perished in World War I – died not as a direct result of injuries sustained from explosions, shrapnel, bullets or poison gas, but of gangrene, tetanus and other infectious diseases. The deadly bacteria entered their bodies through injuries that, in and of themselves, were not necessarily fatal. They settled deep into wounds where the antiseptics of the day could not reach them, rendering those antiseptics practically useless. In addition, they killed white blood cells, reducing the body's natural ability to fight infection. Fleming and his colleagues therefore suggested not treating the wounds with antiseptics, but only flushing out them with salt water. It was really a groundbreaking experiment for the time (and probably would be today as well). Alexander Fleming presented his suggestion at a scientific forum, but again failed to capture the interest of the audience, and so his research was not accepted.

But it was his experience with deadly infections and the powerlessness in the face of the patients' suffering that spurred Alexander Fleming to change the focus of his science career. He decided to search for compounds that were effective against these deadly bacteria.

It did not take long to produce the first results and, as often before, serendipity appeared to help Alexander Fleming. He had a cold in November of 1921, a condition relatively common in London at that time of year. He was studying bacteria colonies in his Petri dishes when a bit of mucus from his nose unexpectedly dropped into one of the dishes. A Petri dish is a shallow glass or plastic laboratory vessel with low edges, something like a cylindrical watch glass, about 10–15 cm in diameter. A culture medium is placed in the bottom of the dish that is then inoculated with the sample of microorganisms to be studied. These multiply and spread in the bottom of the dish to form colonies that can be easily studied.

Fleming got a surprise a few weeks later when he checked on the colonies. There were no bacteria in the spot where the drop of his nasal mucus had landed. Something that was present in the mucus had killed the bacteria. Repeated experiments brought identical results, and the same happened when he used the nasal mucus of all his colleagues. That was not the last of the surprises, however. The bacteria responded similarly to tears, saliva, blood plasma, and other body fluids. He then tested the tears of a variety of animals. Apart from laboratory animals, he also used the tears of horses, cows, pigs (who were allegedly very uncooperative), ducks, and geese, along with about 50 other species of animals living in London's zoo. The results were conclusive: most body fluids must contain a bacteriolytic substance – a substance that dissolves and removes (lysizes) bacteria. The substance was given the name lysozyme.

Lysozyme research opened up a huge area of bacteriology, but it was of little value in practical life because it was not effective against all types of bacteria, only the harmless microorganisms that did not cause any diseases. Furthermore, the research was presented by Alexander Fleming and once again it was given little regard.

But then came that memorable day – Monday, September 3, 1928 – and the greatest serendipitous event of Alexander Fleming's life. On that warm, sunny day, he returned home from his month-long vacation. He and his wife and son had been to their country home in Suffolk, The Dhoon. He had purchased the home on a mortgage seven years earlier and it had since become a favorite place for his entire family to spend their free time. They spent weekends and holidays together there, just as they had vacationed there in that August of 1928, and there was nothing to suggest that this particular summer would be extraordinary in any way. But during his vacation, an incredible chain of fortuitous events took place in a succession that borders on the miraculous. As observed at the start of this chapter, the entire story is still the stuff of myth, legend, and various apocryphal interpretations.

So, what exactly happened that summer? What were these events that led to Alexander Fleming's phenomenal discovery?

Before leaving for vacation, Fleming was studying a gold staphylococcus (*Staphylococcus aureus*). This is a bacterium that causes many diseases, from simple skin conditions to the fatal sepsis. Fleming was an experienced bacteriologist who could tell the age of bacterial colonies by their color and then determine, by their age, how infectious they were. After he finished his research, he did not discard the Petri dish; instead, he set it aside to observe later when he returned from vacation. When he observed the colonies on that fateful Monday, most of them looked as expected, except for one that was contaminated with mold. It would not have been anything unusual, since there were microorganisms everywhere and some could have settled on this particular colony of bacteria. What was unusual, however, was that while the entire dish was thick with bacteria cultures, there were no gold staphylococcus in the vicinity of the mold, just an empty ring. His experience with lysozyme told him that there must be something in the mold that could stop the growth of bacteria. "That's funny," he said out loud. He had just discovered penicillin.

Incidentally, the dish containing the mold and bacteria was dried and preserved and is part of the collection at the British Library.

It was clear from the beginning that the mold produced a substance that destroys bacteria, but it took some time before it was determined how Fleming's discovery had happened. The mold that contaminated the bacteria colony – *Penicillium notatum* – only acts on bacteria that are young and actively multiplying, but the bacteria in Fleming's Petri dish were old. For the mold product to be effective, the mold (or rather the spores) would have to get to the bacteria in their early stages, at the time when Fleming inoculated them with the culture medium. Moreover, the ideal temperature for the growth of gold staphylococci is higher than the ideal temperature for the growth of mold. That means that temperatures in the summer months (Fleming vacationed in August) are more beneficial for bacteria growth, which age faster than the mold can grow. Here is where the first of an entire chain of serendipitous occurrences comes in. According to meteorology records, a low-pressure system and unseasonably low temperatures affected the weather in London for the first nine days of August 1928. The *Penicillium notatum* had plenty of time to multiply, but the growth of the bacteria was slowed by the low temperature. A heat wave hit on August 10, bringing temperatures that were favorable for the growth of bacteria, but they were now growing in an environment full of developed mold capable of killing young bacteria. Had Alexander Fleming left for vacation a week earlier or a week later, or if he had observed the bacteria in any other year, the discovery would never have been made.

Another question was how did the relatively uncommon mold, *Penicillium notatum*, come to be in the Petri dish containing the staphylococci? Various rumors swirled about: it came from a moldy cheese sandwich Fleming had eaten for lunch; it was caused by the bombing of the nearby railway station, which stirred up mold in the area. Unlikely, since the station was bombed during World War II and Fleming most certainly made his discovery in 1928. The pub across the street from Fleming's laboratory was also suspected, which was not a great advertisement for a facility of that type.

That brings us to another of the serendipitous events. A young biologist was studying the connection between mold and asthma in a laboratory below Fleming's. One of those molds was *Penicillium notatum*. The spores had a relatively short journey to reach Alexander Fleming's table and they arrived just in time, when he was inoculating the culture medium with the gold staphylococci bacteria.

Penicillin, as its discoverer would later name the product of mold, still faced a long road before it would become part of the world's heritage. Fleming himself did not see any great value in his discovery at first, and when he stood before professionals of the sector to present his achievements, he was unable to capture their interest. His lecture in 1929 and the publication of his report that same year in a medical journal received little attention. The main reason why penicillin was not initially successful was that at the time it was discovered, there were few multidisciplinary science laboratories – the institutions we know today where chemists, biologists, pharmacologists, and clinicians work together side by side. Fleming was a bacteriologist and had a limited knowledge of chemistry and clinical practice, as well as limited connections among experts in other science disciplines. Penicillin research eventually fell by the wayside and between 1930 and 1940, penicillin was not mentioned in any of his publications.

Fleming had personally only helped one single patient with penicillin. That was in 1932, when young medical student Keith Rogers, a member of the rifle club at St. Mary's Hospital, developed conjunctivitis so severe that he would be unable to take part in the upcoming match. Fleming, who had been an enthusiastic marksman when he was still known by the name Sandy, understood just how his younger colleague felt and decided to help him out. As Rogers later said, "*he put in some yellow fluid, which he assured me was safe and which I imagine was penicillin, that was made in the lab then*." No matter how it actually happened, Keith was cured and able to participate in the match.

Alexander Fleming, Alex, Sandy, and Flem, would become only one of the interesting, if not very successful, characters in the history of medicine. The entire story could have ended there, except that several years later a team of scientists from nearby Oxford, equally as fervent as the discoverer himself, would transform penicillin from a laboratory oddity into a wonder drug.

But that is the subject of the next story.

Story 5.4: Three Englishmen and the benefits of America

Rarely does a group of such diverse people ever get together like the group that gathered in the late 1930s at the Pathology Institute at the University of Oxford. Three completely different individuals met there in one fortunate confluence. They were from three different countries and each was of a different origin. Nevertheless, they worked together to create something that made them famous. What they achieved – penicillin – was only accomplished because each of them made his own unique contribution, applying his own type of genius. That each one of them was indispensable is best summed up by one of their contemporaries: "*Without Fleming, no Chain or Florey; without Florey, no Heatley; without Heatley, no penicillin*." By today's standards, this Oxford team was small, but its achievements were monumental.

The leader of the team, Howard Walter Florey, was from Australia. He was born in 1898 in Adelaide, where his father had a prosperous shoe factory. As a student, he went to England and studied physiology at two renowned universities, Oxford and Cambridge. At the age of 33, he became a professor at the University of Sheffield and four years later, in 1935, became the Head of Department at the Dunn School of Pathology at the University of Oxford. He remained there for 27 years.

Florey was a tall, handsome man with a square chin and dark hair parted down the middle. Despite the amount of time he spent in England, he retained his typical Australian accent. He was usually quiet and reserved, but could be cutting and sarcastic. As an excellent and enthusiastic organizer, in a very short time he had turned his department into a multidisciplinary team composed of mostly young, equally enthusiastic scientists.

One of these scientists was a biochemist named Ernst Boris Chain. He was born in 1906 in Berlin into a Jewish family of Russian and German descent. In 1933, shortly after he graduated from university in Berlin with a degree in physiology and chemistry, Adolf Hitler rose to power. The young Chain quickly understood that with his family heritage, he would face problems in his homeland. In April that same year, he left Germany with £10 in his pocket and went to London. He later said of himself that he was one of the first scientists to emigrate from Germany to England. He joined Howard Florey's team in 1937 and was the exact opposite of his boss. He was short, mercurial, and quick-tempered, with black hair and a black moustache, similar to the brilliant Albert Einstein. He also had a love of music in common with the great physicist. While Einstein was a talented violin player, Chain could have chosen between a career as a biochemist or a concert pianist.

The third member of Howard Florey's group, another biochemist, was Norman George Heatley. He was born in 1911 in the county of Suffolk, East Anglia, to the family of a veterinarian. He graduated from Cambridge with a degree in the natural sciences. About three years later, Howard Florey invited this sharp, skilled chemist to join the group in the pathology department. He was the antithesis of Ernst Chain, quiet and modest, concealing his scientific prowess behind his mild-mannered and unobtrusive behavior. The contrast between the mild-mannered Heatley and the quick-tempered Chain was legendary and much effort was required on the part of their boss to keep things under control.

The seeds of their interest in penicillin were planted back in 1938, when Chain came into possession of the long-forgotten research Alexander Fleming had conducted in 1929. This was the research in which Fleming described his first experience with the action of mold on bacteria. The paper captured their interest, and when a grant came in from the Rockefeller Foundation, the research could begin.

The first problem was getting enough penicillin from the mold to perform some standard testing of its properties. Alexander Fleming's original materials included two units of penicillin per milliliter of solution. The average daily dosage of penicillin is approximately 15,000,000 units, so if we wanted to get a one-day treatment of the penicillin contained in Fleming's Petri dishes, we would need several football fields worth. The mold *Penicillium notatum* needs oxygen, so a sufficient air supply must be ensured. It could be cultivated in the standard way, by inoculating the surface with a culture medium and waiting for a few days for the penicillin to grow to the maximum quantity in the medium, then repeating the entire procedure with a new culture medium and a new cultivation. Heatley, however, discovered that the mold continues to grow and can produce more penicillin if the culture medium is constantly replaced. They tried cultivation in all kinds of cylindrical vessels in which this procedure could be conducted. They tried shallow bottles of various sizes, empty cake tins, gas cans, and also bedpans – which proved to be the best-suited for the job. Mold could be cultivated in the bedpans on the surface of a culture solution that could be constantly replaced. Heatley designed ceramic vessels of a similar shape, and a friend of Howard Florey's who owned a ceramics factory manufactured about 600 bedpans according to Heatley's drawings.

The pathology department turned into a cultivation factory. When Heatley also developed a more effective method for obtaining penicillin from mold, Howard Florey's team already had about 100 mg of the brown powder. It contained 1,000 times more penicillin than that contained in the original yellow solution. It was huge progress, but the majority of the material was still made up of contaminants. Penicillin only comprised one hundredth of a percent, but that was enough to conduct the first experiments on animals.

Saturday, May 25, 1940, was the day on which they infected eight white laboratory mice with a deadly dose of streptococcus. Four of the mice were then given a solution of the dark brown powder containing penicillin. The other four were left without treatment as a control group. By Sunday morning, the mice in the control group were dead but all four that had received penicillin were alive. We already know how the three scientists reacted; their responses were as different as their personalities. Howard Florey very modestly said, "*it looks quite promising,*" Ernst Chain danced around the laboratory with joy, and Norman Heatley said the success was because he had put his underpants on back to front the day before.

Serendipity also played a role in the first *in vivo* success of penicillin. In fact, serendipity helped the trio of experimenters out twice. The first time was in the selection of laboratory animals. We saw how important the right (and random) selection could be in one of the stories about vitamin C; Howard Florey's team had similar luck to Axel Holst's team. Had they selected guinea pigs instead of mice, they would have failed because penicillin is toxic in guinea pigs. The experiment would have resulted in eight dead lab animals, both the control group and the treated group, and the research would probably have been abandoned.

The second serendipitous event was that not one of the mixtures containing the brown powder was toxic in the mice. Considering how contaminated it was, it is quite extraordinary that it was not. If even just one of the mixtures had been toxic, it would have caused the death of all the test subjects – the treated ones as well as the untreated ones.

Since the trial on mice had been successful, Howard Florey decided to take a revolutionary step and administer penicillin to a human patient. By then their preparation was more purified, although it still only had a penicillin content of three percent. The first patient was 43-year-old Oxford policeman Albert Alexander. He had a scratch on his cheek from a rose thorn, and although the cut was tiny it had caused a severe streptococcal and staphylococcal infection. Four months after receiving the cut, his body was covered in festering sores, he had lost an eye, and had a bone marrow infection. The prognosis was bleak.

The first dose of penicillin was administered to the patient on February 12, 1941, but unfortunately the research team from the pathology department did not have enough of it to cure him. The body excretes a large percentage of penicillin in its original state in the urine, which provided another source of penicillin. However, that meant taking the patient's urine, getting on a bicycle, and racing from the clinic where the patient was being treated to the pathology department. There, the penicillin could be extracted from the urine and then sent back to the clinic. They had to go back and forth several times. Unfortunately, not even these heroics could save the patient because there was simply not enough penicillin. Although his condition improved after five days, Alexander relapsed and died a month later.

It was clear after this failure with the first patient that without enough penicillin, their efforts were pointless. They also knew that the modestly-equipped pathology department, with its 600 ceramic bedpans, would not be enough. They needed industrial-scale production.

The first attempts at the mass production of penicillin in Great Britain were less than brilliant. They had no experience, and to make things even worse, there was a war going on. It was the time of the Blitz. London was being bombed and there were more urgent priorities than cultivating mold. The main problem persisted – the mold only grew on the surface of the culture media. Although the efficiency of penicillin production had improved since Alexander Fleming's time, a huge surface area was still required to produce the desired quantity of penicillin. They also had to solve the problem with the fermentation vessels for cultivating mold. The original ceramic bedpans designed by Norman Heatley were fairly expensive to make. There were numerous obstacles, but the large food and pharmaceutical companies in Great Britain did their best to deal with them. Glaxo, originally a manufacturer of powdered milk and vitamins, used glass bottles to cultivate mold. They built a plant large enough to house 300,000 of the bottles. Boots went even further, with a plant that could hold a million bottles. It was impossible to manufacture such a huge quantity of special glass immediately, so they used milk bottles. This caused a shortage of glass bottles all over England. The British Royal Navy manufactured its own penicillin, using gin bottles instead of milk bottles. The gin bottles were cylindrical and the navy had plenty of them.

Regardless of the cultivation vessels, penicillin manufacturing in Great Britain grew by leaps and bounds. In 1943, monthly production was at 300,000,000 units, and a year later it was more than three billion. By 1945, production was at nearly 30 billion units.

The situation was quite different on the other side of the Atlantic. When Howard Florey saw that penicillin manufacturing in Great Britain was off to a slow start, he decided to try his luck in America. He and Norman Heatley went to the U.S. in 1941, leaving the other biochemist out of the picture. They kept the preparations for their trip a secret from Chain for many months, and he only found out about it barely 30 minutes before their flight, when he noticed the luggage in the corner of the room. It infuriated him.

The trip to the U.S. was more than dramatic. It was wartime and there was no direct – or safe – connection between London and New York. First, they flew in a blacked-out airplane from Bristol to Lisbon, and from neutral Portugal through the Azores and Bermuda to arrive in New York. It took them exactly one week. The entire time, they had to take care of their precious culture they had taken with them: *Penicillium notatum*.

They were welcomed in the United States with great respect and they did not have to try too hard to persuade the proper officials of the benefits of working with them on the production of penicillin. The little known Northern Regional Research Laboratory was recommended to them. It was located in Peoria, Illinois, about 165 miles from Chicago and relatively far away from the prominent research institutions. At the time, Illinois was one of the largest corn growers in the United States, and the laboratory was used for researching alternative uses for corn. One of the not-very-alternative uses was distilling whiskey. Peoria was one of the largest producers in the world, earning the nickname "Whiskeytown."

Before long, the decision to cultivate massive quantities of mold in Peoria proved to be the right decision. The laboratory was state-of-the-art and had a team with vast experience of growing mold. They used a method that was not yet known in Great Britain: instead of using the surface of the culture medium, they used what they called deep-tank fermentation. The mold did not grow *on* a liquid medium but rather *in* it and it was constantly agitated and aerated to make sure there was enough oxygen. The laboratory used tanks containing over 500 gallons of the culture medium for the deep-tank fermentation. The first big contribution of the laboratory in Peoria was that an enormous amount of space was saved and the entire penicillin production process was streamlined.

While the first steps towards the mass production of penicillin were led by the experience and smarts of the research team, the next two steps were again helped along by serendipity. One of the waste products from corn processing was corn steep liquor, and for years they did not know what to do with it – until they tried to use it as a fermentation medium completely by chance. It was a success. The liquor not only made production cheaper, it also resulted in 12 times the amount of penicillin from the mold than with the fermentation media that had been used previously.

Even so, the amount of penicillin produced by the mold was still not enough to begin mass production. Another species of mold had to be found, one that would produce more penicillin. The laboratory sought the help of the United States Army. Mold samples from around the world soon began arriving in Peoria, and serendipity once again played its role. The mold containing the greatest amount of penicillin did not come from a far-away country but was discovered by laboratory assistant Mary Hunt on a moldy melon at a market in Peoria. The penicillin contained in this mold was 100 times the amount contained in the original species brought over from England.

There was nothing now to stop them from starting the mass production of penicillin in the United States. The first company to enter the arena was Pfizer. They began production slowly and carefully, and the first batches of penicillin were made in 52-gallon tanks. But then, in 1943, Pfizer built 14 much larger 7,500-gallon tanks in an old ice factory in Brooklyn. Soon, other manufacturers joined in and penicillin production in the U.S. skyrocketed. In 1943, production was still at about the level of Great Britain; in 1944 it was 40 times greater, and by 1945, an incredible 68 trillion (68,000,000,000,000) units were produced.

Since the production of penicillin began during World War II, the military naturally had the greatest interest in the product. The amount of penicillin available in 1943 was enough to start the first big clinical trials, which began in April 1943 in a military hospital in Utah. The results were positive, and more and more hospitals began to take part. At first, the trials, as well as all practical use, were reserved mainly for the military. In 1943, the U.S. Army received 85 percent of the penicillin that was produced, but production grew at a breakneck rate, and in March 1945, penicillin was made available to civilians.

The story draws to a close here. After World War II, penicillin therapy spread throughout the entire world and it was administered to a growing number of patients. The age of penicillin had begun, and together with the technological revolution in medicine, the fear of infection became a thing of the past for the post-war generation.

And what of the heroes of this story? After the war, the reserved Howard Florey remained as director of the School of Pathology at Oxford until 1962. The impulsive Ernst Chain broke off from his boss and went to work in Rome. Norman Heatley also considered leaving Oxford, but Florey persuaded him to stay on and work with him. Howard Florey and Ernst Chain were knighted, and in 1945 they received the Nobel Prize in Medicine along with Alexander Fleming. The world seemed to forget about Norman Heatley. Towards the end of his life, when his health was starting to fail, Heatley passed the time making miniature toys from bird feathers.

Story 5.5: Stubborn Andy and the need for meat

It very well may have all started back in 1943, when Howard Florey had a little leftover recycled penicillin from one patient's urine. The therapeutic effect of penicillin in animals had been known ever since that fateful experiment on eight mice in May 1941, which was why Florey offered it to a veterinarian friend to try on slightly larger animals. There was enough penicillin extract for two cows, so the veterinarian attempted to use it to treat their mastitis, an inflammation of the udder tissue. The attempts were successful. This opened up more possibilities for the use of penicillin and other antibiotics and they were put to use in veterinary medicine as well as human medicine.

Due to this veterinary use, the production of meat, milk, eggs and other food was significantly increased after World War II. Antibiotics contributed to a revolution in both medicine and agriculture. They played a huge role in eliminating two of the greatest human fears: disease and hunger.

In the 1950s, meat production gradually transitioned from small-scale farming to industrial agriculture. Enormous herds of cattle were housed in giant barns, and the care of animals became mechanized and automated. Feeding, milking, and manure removal were all handled by machines that today are fully automated. Between 1940 and 1980, the number of farms in the U.S. fell by 80 percent while the number of animals doubled. Pig and chicken farming experienced the same growth. Between 1953 and 1960, the number of broilers increased ninefold. In the author's home country of Slovakia, the number of cattle doubled between 1950 and 1990, while the number of pigs increased fivefold. This increase meant that Slovaks could eat three times more meat.

However, the growing number of animals and the increased concentration of animals in one area, coupled with their lack of physical movement, took its toll in a higher incidence of diseases that spread like wildfire through large herds. Without antibiotics, most of the animals would have died, there would be no large-scale farming, and people would not be able to consume as much meat, milk or eggs. Antibiotics became an inseparable part of the diet of farm animals.

Apart from the expected therapeutic effect, antibiotics had another wondrous benefit. They helped the animals grow faster. This antibiotic property was discovered completely by accident in the late 1940s, but the consequences were far-reaching. Just a spoonful of pure penicillin per ton of live weight – a much smaller amount than was needed to kill bacteria – and the animals grew at a 5–10 percent faster rate. In addition, they needed one-tenth less food for that growth. The administration of a small amount of antibiotics significantly reduced the costs of meat production, but on the other hand, it made pathogenic bacteria stronger rather than weaker.

At first, antibiotics in farming did not present a serious problem for humans, or so it was thought. Infectious diseases in humans and animals are usually caused by different strains of bacteria. A small amount of antibiotics in animals could create treatment-resistant bacteria, but only in those animals. In theory, it should have posed no risk with regard to resistance to pathogenic microorganisms that cause infectious diseases in humans. Unfortunately, this theoretical assumption turned out to be wrong. Reports soon began to appear about the negative effects of antibiotics in food on human health, but there was no conclusive evidence, and since increasingly more meat, milk, and eggs were needed, the use of antibiotics in farming continued. But we cannot be too critical. The times of hunger were still a fresh memory in the minds of the post-war generation, and the need for enough food outweighed the fear of any potential danger posed by the overuse of antibiotics.

The first problems started with milk. The mass production of milk required mechanical milking, which increased the incidence of the mammary gland infection known as mastitis. The fact that the annual losses incurred by the industrial farms in the U.S. due to the disease ranged from $1.7 to $2 billion shows just how serious and common the disease was. The simplest and cheapest solution to the problem was antibiotics, and their use became widespread. There was just one little catch. Penicillin reliably gets rid of infections in a fairly short amount of time, but it takes somewhat longer for it to be excreted from the system – and that means it gets into the milk.

The first signs that there was something wrong with the milk began to appear in the early 1950s. The milk did not sour and no cheeses could be made from it. Certain strains of bacteria are required to sour milk, and bacteria are also required to make cheese. Where there is penicillin, there are no bacteria.

This "sanitizing" of the milk was not the main problem associated with penicillin, however. The long-term consumption of small amounts of antibiotics in milk made people hypersensitive to antibiotics, which resulted in allergic skin reactions. That was when the wonder drug began turning into an undesired admixture. In the 1960s, the fear of allergies practically changed a useful and healthy beverage into a poison liquid. Milk has still not fully recovered from the downfall.

Reports on the dangers of antibiotic overuse in animals leading to resistant bacteria in humans were on the rise. It was only a matter of time before the scales were tipped and the problem of resistant bacteria would outweigh the benefits of the use (and mainly the overuse) of antibiotics in farm animals.

In the 1960s and 1970s, numerous research scientists were looking to tip those scales, the most notable being the English bacteriologist Ephraim Anderson.

One of his colleagues said that working with Ephraim Saul Anderson was akin to driving a car without springs across a lava field. Andy, as his colleagues called him, was a fighter and he was tough on himself and those around him. He was not afraid of conflict and was not always tactful. Sometimes he was downright abrasive. He had acquired these characteristics in childhood, which he spent in a working-class neighborhood in the industrial town of Newcastle-upon-Tyne in the northeast of England, where he was born in 1911 into a family of Jewish immigrants from Estonia. But he was gifted and hardworking, graduating from medical school at the age of 22. He gained medical experience as a practical and military physician, and in 1947 he went to work for the Enteric Reference Laboratory in north London. He became its director in 1954 and remained in that position until he retired in 1978.

Anderson got into his first conflict in 1960, having developed a method of optical microscopy that allowed viruses to be seen. It was something unprecedented at the time, as it had only been possible to see viruses using the substantially more difficult electron microscopy technique. His method was harshly criticized by experts in microbiology, including a Nobel laureate, and he fought back in his own way, which was not always diplomatic. He was finally given recognition in 1997 – 20 years after his retirement – when one of his images of a virus made the cover of a respected journal.

Several years later, he was involved in another conflict. It was still the prevailing opinion at the time that animal and human pathogenic microorganisms were different in most cases. Even when a new strain of bacteria was identified that was resistant to treatment, due to the overuse of antibiotics, it only concerned animals and had no substantial impact on humans. Andy Anderson proved in his laboratory that the genes of antibiotic-resistant microorganism strains could be transferred from animal bacteria to bacteria that caused deadly diseases in humans. Resistance to antibiotics, and the resulting inability to treat infectious diseases, could set humankind back 30 years.

It was a powerful warning, but Andy once again faced opposition. This time it came from bureaucracy. There was an unwillingness to change the views on administering antibiotics to farm animals. Committees were formed and reports were issued, but the situation remained unchanged. However, evidence was mounting that Anderson was right. In 1967, several children in Middlesbrough became ill with gastroenteritis – an infection of the stomach and intestines. This illness normally responded to antibiotic treatment, but in this case, it was resistant and the children died from what was a common infection. There was a much worse epidemic in Mexico several years later, when over 10,000 people contracted typhoid fever. The bacteria that cause this disease were resistant to practically every primary antibiotic.

Anderson continued to wage his campaign and continued to make enemies. A committee was finally formed in Great Britain in 1968 that was supposed to assess in depth the impact on human health of the antibiotics used in veterinary practice. The agriculture ministry was emphatically against Andy holding a position on this committee, but they ultimately came to a decision that, while not completely eliminating the use of antibiotics in animals, at least reduced their use. They made a distinction between antibiotics for therapeutic use and the use of small doses for promotion of growth. The first use was kept and they recommended banning the second use. The British government accepted the committee's recommendation and banned the use of antibiotics as growth promoters. Other countries followed Great Britain's lead, with Czechoslovakia implementing the decision in the early 1970s. Although a small step, it was significant for penicillin. It meant that its use in human medicine was no longer threatened by its routine administration to animals as part of their feed.

The overuse and abuse of antibiotics – in people as well as animals – did not come to an end and more or less continues today. In 2015, antibiotic resistance was responsible for the deaths of 23,000 people in the United States and 2,000 more than that in Europe. Antibiotic abuse does not only concern farmers, it also concerns doctors, pharmacists, manufacturers, and regulatory bodies. Not to mention us, as its consumers.

Andy Anderson's battle ended in 2006. When he retired, he was finally able to enjoy all of his other interests, like music, history, travel, and the amateur manufacturing of perfumes that he kept in antique bottles.

Story 5.6: The renowned health professional and ethical blindness

This story only marginally touches on the history of penicillin, but it carries a powerful ethical message that makes it worthy of a spot in this chapter. It is actually two stories about two clinical studies. The first took place from 1932 to 1972 in Macon County, Alabama, in the town of Tuskegee. The second study did not last as long – from 1946 to 1948 – and took place in the Central American country of Guatemala. The common denominators were a disease and the doctor who worked on both studies. The disease was called syphilis and the doctor's name was John Cutler.

Tuskegee, the center of the first story, was founded in 1833. At the time the story takes place, the town had a population of a few thousand; the population is not much higher than that today. It is home to one of the best historically black colleges in the United States. In the 19th century, cotton plantations were predominant in the area and they were strongly dependent on slave labor. After the slaves were freed, many stayed to work on farms as sharecroppers, giving a share of their crops as a rental payment. African-Americans make up 96 percent of the population of Tuskegee and it is not surprising that the town played an important role in the history of Black America.

However, it made history as the place where an extensive, long-term medical study took place, which had a long – and by today's standards, politically incorrect – name: *The Tuskegee Study of Untreated Syphilis in the Negro Male*, more widely known under the shortened name of *The Tuskegee Syphilis Experiment*.

There were several reasons why this town of Tuskegee was chosen for the study, but the main reason was that the southern United States and its black population had a significant problem with syphilis in the early 20th century. In Macon County, 35 percent of African-Americans had contracted this sexually transmitted disease. In the early 20th century in Tuskegee, an educational program was begun aimed at the advancement of the black population, which included improving the health of the population – and syphilis was definitely a part of the problem. There was a fair amount of funding for this project, provided mainly by philanthropic foundations located in the northern U.S.

Everything started off with good intentions. The original goal was to do a brief, six-to-nine-month study of untreated syphilis, after which available treatment would be continued. At the time, doctors could choose from three different treatment methods. They could treat patients with toxic mercury or the less toxic bismuth, or with Paul Ehrlich's "magic bullets" – Salvarsan and Neosalvarsan. These methods were not always effective and there were severe side effects, but they did save the lives of many patients.

The study was officially organized by the U.S. Public Health Service in cooperation with Tuskegee University. There were 600 black men enrolled in the study, of which 399 had been diagnosed with syphilis and the remaining served as the control group. They were all impoverished persons, mostly illiterate sharecroppers, the sons and grandsons of former slaves. They were promised benefits for participating in the study which most blacks could only dream of at the time: free medical care, transportation to and from the clinic, free meals during examinations, treatment for minor illnesses, and even money for their funeral.

Before the study could get off the ground in its original form, the stock market crashed in 1929. Sponsors withdrew funding and there was no money for the study. The organizers decided to go ahead with the study anyway, but with one difference: the patients did not receive treatment. For 40 whole years.

Ethical standards were violated from the beginning. Participants were promised benefits, but they were not informed that they had syphilis. They were not even told they were in a study of the disease. All they knew was that they had "bad blood." They were grateful they were being cared for and they willingly underwent examinations. It lasted many long years. The experiment was not a secret, with reports regularly published in medical journals. A total of 127 black medical students rotated through the study as part of their medical education.

Meanwhile, the world was making progress in the battle against bacterial infections. In 1935, Prontosil became available, then additional sulfonamides came on the market, while penicillin had been available since 1945. In 1943, the Henderson Act was passed in the U.S., which required mandatory diagnosis and treatment for venereal diseases.

The end of World War II brought shocking revelations about the practices of the Nazi doctors and their experiments on humans. After the verdicts fell at the Nuremberg trials against the main participants in the medical experiments, the Nuremberg Code was announced in 1947. In ten points, it defined the rights of research subjects and included principles such as informed consent of the participants, the absence of coercion, clear formulation of the scientific goals and consequences of the study. In 1964, the World Health Organization (WHO) created a document based on the Nuremberg Code that clearly set out the rights of research participants. It was called the Declaration of Helsinki, based on where it was drawn up, and is still considered a milestone in the ethics of human research.

None of that ever crossed the imaginary border of the Tuskegee experiment. The organizers of the study, including the U.S. Public Health Service, intentionally withheld information from the participants about treatment, which at the time was readily available. Not only that, but when the war ended, the United States government sponsored various public campaigns to eradicate venereal disease. None of those campaigns ever found their way to Tuskegee.

The Tuskegee experiment did not exist in a vacuum, however, and its organizers were warned on several occasions about serious breaches of basic ethical principles. They managed to defend themselves for a long time; their goal was to end the study when the last of the participants died. Their arguments were supported by both the American Medical Association and the National Medical Association, the latter an association of African-American physicians.

When the study finally ended in 1972, only 74 of the participants were still living. Of the original 399 infected men, 28 had died of syphilis and 100 had died of related complications. About 90 were able to withdraw from the study and get adequate treatment, but 40 wives were infected and 19 children were born with congenital syphilis.

Everything was made public after the details of the study appeared in the headlines of the leading American newspaper, *The New York Times*, on July 26, 1972. The revelations were appalling. Congressional hearings concerning the Tuskegee experiment were called immediately, and then a committee was formed to look into the experiment. The outcome was unequivocal: the scientific protocol of the study had been grossly breached to ignore the safety and health of the participants. According to the committee, the study was unethical and was to be terminated immediately. It was terminated in November 1972, and a year later the National Association for the Advancement of Colored People filed a class action lawsuit against the U.S. government. In 1974, the government undertook to pay nine million dollar in damages to all the survivors and their descendants and to pay for all their medical care.

It was another 43 years before the United States officially apologized for the study. That apology came through the words of then-president Bill Clinton: "*The people who ran the study at Tuskegee diminished the stature of man by abandoning the most basic ethical precepts. They forgot their pledge to heal and repair. (…) Today, all we can do is apologize.*"

The last participant in the Tuskegee experiment, Ernest Hendon, died on January 25, 2004 at the age of 96.

Reflections on this unethical, 40-year human experiment have lasted till today. How was it possible that not one of the organizers doubted the ethics and moral value of the experiment? No doubts were raised by local organizers in the African-American community in Tuskegee or by the African-Americans who had an active role in the experiment. The highest medical authority in the land – the U.S. Public Health Service – oversaw the experiment the entire time. They were all convinced they were serving a good cause.

No doubts were raised by John Cutler, a senior surgeon for the U.S. Public Health Service, who worked on the experiment in the 1960s. He not only worked on it, he vehemently defended it after it was shown to be immoral. Before we get into this controversial doctor, let us look at another experiment on human volunteers. This one concerned the same disease as the Tuskegee syphilis experiment. It was called the *Guatemala Syphilis Experiment* and not only did John Cutler work on this one, he actually organized it.

The experiment took place from 1946 to 1948 and involved various official institutions. On the American side, it was the already mentioned U.S. Public Health Service, supported by a grant from the National Institutes of Health, and the Pan American Sanitary Bureau. On the Guatemalan side, it was various state medical, military, and correctional organizations.

The experiment consisted of infecting the research subjects with sexually transmitted diseases, mainly syphilis and gonorrhea. The research subjects were not laboratory animals, they were Guatemalan prisoners, soldiers, and mental patients. A total of 1,308 people, aged between ten and 72 were infected. To infect the prisoners and soldiers, they used sex workers that had already been intentionally infected. For the mental patients from the National Psychiatric Institute, they used artificial methods to infect them, placing infected materials on their reproductive organs and mouths, or injecting it under their skin or directly into their cerebrospinal fluid. Artificial methods of infection were also used on soldiers and prisoners if the natural method of transmission failed. The infected persons were then clinically observed and regular serological and other screenings were conducted to monitor the progress of the disease. None of the participants in the study were informed in advance of the purpose and goals of the experiment. There was one positive difference between the Guatemala experiment and the Tuskegee experiment, in that the patients were given penicillin to determine how effective it was against venereal disease. The study officially ended in 1948, but no official report was ever issued. One of the reasons given for ending the study was the high cost of penicillin.

It would be another 57 years before information accidentally came to light. It was discovered while the details of the Tuskegee experiment were being reviewed in the archives of John Cutler, the young doctor who organized and ran the Guatemala experiment at the age of 30.

Now we come back to the doctor whose colleagues regarded him as an eminent medical officer and recognized scientist.

According to his official biography, John Charles Cutler was born on June 29, 1915 in Cleveland. In 1941, he graduated from medical school in his hometown and took a job right away with the U.S. Public Health Service. He specialized in venereal disease, and in 1943 he went to work as a medical officer in the U.S. Public Health Venereal Disease Research Laboratory. In 1948, the WHO tasked him with leading the fight against venereal disease in India and in 1958, he became the assistant surgeon general of the U.S. In 1967, he was named professor of international health at the University of Pittsburgh. He was the chairman of the healthcare department, and in 1968–1969 he served as the dean of the School of Public Health. When he died in February 2003, the obituaries contained sentences such as "*He thought every person should have access to these services (healthcare), regardless of income*," or "*To him, health was more than simply studying microbes. It was life*."

The irony is that we are talking about a man who wanted everyone to have access to healthcare services, yet denied those services to dozens of his black fellow Americans in the South, in Tuskegee. This man, for whom health was more than just the study of microbes, was capable of infecting healthy people with those microbes. A man who, in 1949, right after the Guatemala experiment ended and long before he participated in the Tuskegee experiment, coauthored a publication fervently promoting penicillin as a treatment for syphilis. John Charles Cutler can serve as an example of a person whose priority is scientific achievement, a priority that is so inherent and important that he cannot make any room for the ethical and moral aspects of his research. This is usually called ethical blindness. Cutler and his colleagues were definitely not inhuman monsters and they believed that they were using science to fight a deadly disease. Cutler thought that the Tuskegee study was "*…grossly misunderstood and misrepresented… these individuals were actually contributing to the work towards the improvement of the health of the black community rather than simply serving as merely guinea pigs…*".

The experiments on humans in the United States cannot be compared to the Nazi experiments conducted during World War II. However, the Tuskegee and Guatemala experiments do have one thing in common with those conducted in Germany: the supremacy of the experimenters over the participants. The experiments conducted in the concentration camps were clearly about the supremacy of the so-called Aryan race. For the experiments in the American South and in Guatemala, skin color played a role. Just as President Bill Clinton apologized to the victims of the Tuskegee experiment, his wife Hillary Clinton expressed her regret about the experiments conducted in Guatemala. In 2010, as secretary of state, she apologized to the victims of the experiment with these words: "*We deeply regret that it happened, and we apologize to all the individuals who were affected by such abhorrent research practices*."

Concluding remarks

We began the stories about penicillin with the famous return, surrounded by myths, of Alexander Fleming from vacation. Even though the real story of penicillin is far more complicated and involves numerous people, this one man remains at the forefront. Every educated person knows his name and connects it with his discovery. Sir Alexander Fleming is a name that is familiar in every society and his discovery of penicillin is probably the most famous of all drug discoveries. Penicillin changed the destiny of humanity. We no longer need to fear infectious diseases or infections caused by an ordinary scratch from the thorn of a rose. In the 90 years since Fleming's discovery, dozens of antibiotics have been described and used, but it was penicillin that readers of *The Times* voted as a treasure of humanity in 2000. It is produced and consumed in greater and greater quantities today, with worldwide sales of penicillin and similar antibiotics at somewhere around $15 billion. In spite of this, the price keeps coming down. In the ten years from 2000 to 2010, penicillin production increased by 30 percent.

But there are still challenges facing penicillin and other antibiotics. First of all, there is a problem with availability in some parts of the world. Participants in the Tuskegee study were definitely not the last people who did not receive penicillin treatment. In 2013, pneumonia cost the lives of 935,000 children under the age of five; if they had had penicillin, most of them would have lived.

With the growing population, there is an increasing need for animal protein, and since antibiotics play a significant role in their production, the consumption of antibiotics is on the rise as well. In 2010, over 60,000 tons of antibiotics were used livestock farming – more than was used in human medicine. Human consumption itself is twice what is really needed to treat infections that can be cured with antibiotics.

Alexander Fleming does not have to worry about these things anymore. Maybe it really was a combination of luck and the ability to observe unexpected occurrences, but the fact remains that he became the most famous doctor in history and one of the greatest heroes in medicine.

6

The Pill

The human reproductive process is pretty complicated. It requires two individuals of the opposite gender, a plan for how those two individuals will find one another, an algorithm to determine if they are right for each other, and finally a place and time when they will do the actual deed. It also requires finding a way for one piece of genetic information to connect with another piece, and then a way for those two pieces of information to connect to produce a completely new individual. It is quite a complicated and ineffective way to transmit genetic information. It is not really our fault, though; we inherited this process from our extinct animal ancestors. We have accepted it, however, and we do not pay much attention to how complicated it is.

There is one significant difference between us and our extinct animal ancestors: we enjoy the act of reproducing. Humans are the most sexual beings in the animal kingdom, and officially the only beings that take pleasure in procreation. In his book *The Naked Ape*, a bestseller in the 1960s, Desmond Morris explained that our sexual pleasure was the result of the need to develop and maintain a strong pair bond between a man and a woman. Whatever the reason, procreation is definitely not the only goal of the sex act.

There have always been attempts at separating pleasure from procreation, whether it was for moral, medical or social reasons. The real history of contraception, in the truest sense of the word, is limited to the last 150 years; in fact the word *contraception* is less than 100 years old. The need for contraception as a systemic approach arose when women realized that making and raising babies did not have to be their one and only destiny. Unlike the previous chapters, in these stories, in this one women play the key role.

V. Marko, *From Aspirin to Viagra*, Springer Praxis Books,
https://doi.org/10.1007/978-3-030-44286-6_6

Story 6.1: Madame Restell and Fifth Avenue abortions

A brownstone mansion stood on the corner of Fifth Avenue and 52nd Street in the late 1870s, at the spot where a Salvatore Ferragamo boutique store is now located. It was four stories tall, with six large windows on three of the four stories. The entrance was above street level, a wide staircase adorned with a stone bannister leading to the front door. It was a lavish mansion, owned by a stern-looking woman of around 60 years of age who was always richly dressed. She could be seen every morning as she purposefully strode down the steps and climbed into a shiny carriage drawn by two horses. Although you may not know it at first glance, this woman was one of the most famous people of her time in New York and her name often appeared in the tabloids. She was a real celebrity but had an unenviable reputation. She was known as Madame Restell, and the people of New York had given her the title *The Wickedest Woman in New York*. Her luxurious mansion fared no better and was known as "*the mansion built on baby skulls*."

Madame Restell was an abortionist and her mansion on Fifth Avenue was where pregnant women came seeking help when they would not or could not give birth to a child. In New York, she was not the only "female physician," as her profession was euphemistically known, but she was certainly the best known. Which also made her the richest.

Madame Restell was far from being the first woman to provide abortions. The history of attempts to remove an unwanted fetus forcefully before birth is as long as human history itself. The list of the means and devices used for this purpose is pretty frightening. Primitive peoples used all sorts of physical activity, like strenuous labor, lifting weights, and even falling down stairs and diving into cold water. Other methods included starvation, bloodletting, sitting on hot receptacles, and tightly binding the abdomen. The ancient world adopted all these methods and added a few more, such as diuretics, horseback riding, and enemas. The ancient Greek physician Soranos recommended something called the "Lacedaemonian Leap," where the woman would jump and let her heels touch her buttocks. This type of exacting physical exercise allegedly was helpful in terminating pregnancy. When nothing else worked, they relied on sucking pumps, curettes, and hooks.

The list of natural substances they used to terminate pregnancy is also rather long. If an alphabetic list of botanicals that were thought to be abortifacient is created, it would start with birthwort and continue with black or green hellebore, catnip, cyperus, Cyprus oregano, dye's madder, hyssop, juniper (until fairly recently, drinking gin in combination with a hot bath was recommended), lavender, male fern, marjoram, pennyroyal, sage, savory, soapwort, tansy, thyme, and watercress. Even that would probably not be the complete list. They used the entire plant, or the leaves, flowers or roots. They made teas with them and mixed them into milk, beer or wine. In addition to plants, other substances deemed to be abortifacient included iron sulphate, opium, and turpentine, not to mention Spanish fly, ants, camel saliva, and deer hairs mixed with bear fat.

The majority of the remedies against unwanted pregnancy were known to Ann Lohman, the real name of Madame Restell. Born on May 6, 1812 in Painswick, England, she married Henry Summers at the age of 16 and together they emigrated to New York in 1831. Her husband died soon thereafter of bilious fever, and Ann was forced to support herself as a seamstress. Her break in life came in 1836, when she met the educated and cultured Charles Lohman. It was probably Charles who started her stellar career. He invented a story that Ann had trained as a midwife with her aunt in France, whose name she had taken as well. She became Madame Restell, the name under which the couple ran their business. They advertised in New York newspapers, offering women advice and assistance to "*prevent their families from increasing beyond what their happiness would dictate.*" Clients could choose between "preventative powders" and "female monthly pills." The powder cost $5 for a pack and the pills were a little more expensive at $1 each. These were high prices, in today's money $150 and $30 respectively. When clients could not come in person, Madame Restell sent her guaranteed remedies by mail. Her preparations were no different from similar products offered by competitors – essentially a blend of herbs – most often pennyroyal, juniper, tansy, mallow, wormwood, and others. When these products failed, she offered "surgical abortions." She charged poor women $20 for the procedure and rich women five times as much. As her business grew, Madame Restell expanded her services. Women with unwanted pregnancies could stay in her boardinghouse and give birth anonymously. For an additional fee, she offered to broker the adoption of the unwanted babies. Business was soon booming and the Lohmans were able to buy a large lot near Fifth Avenue, on which they constructed their four-story mansion "built on baby skulls."

Early on in her practice, Ann Lohman was not in the habit of breaking the laws of the state of New York. The laws reflected a folk wisdom that a fetus was not a person until the day it first moved inside the womb. An abortion performed before that day was legal, but if she performed one after that day, the "female physician" risked one year in prison. Although it was a legal activity over a certain span of time, public opinion in puritanical America was most definitely disapproving. The standing of Madame Restell and her colleagues was on the fringes of society.

Madame Restell knew the boundaries of her profession and she successfully avoided performing late-term abortions, but she could not avoid publicity. Her name first appeared in the tabloid headlines in 1840 in connection with the death of young Maria Purdy. As Maria lay dying of tuberculosis, she confessed to her husband that the year before she had visited the office of the well-known abortionist and went through an abortion for $20. When her husband heard this, he immediately pressed charges against the "female physician." Although she was ultimately found not guilty, it kicked off a massive campaign that put Madame Restell and her practice firmly in the spotlight of public opinion. They were merciless. The tabloids called her "*the monster in human shape*" responsible for "*the most hellish act ever perpetrated in a Christian land*." They accused her of encouraging prostitution and the irresponsible behavior of young women. She was a hag who preyed on human misery.

That was only the beginning of the tabloid crusade and their twisting of public opinion against Madame Restell. Her name gradually became synonymous with immorality and the term "Restellism" became synonymous with abortion. She was even maligned when she attended a normal birth and assisted with the subsequent adoption. When one of the mothers changed her mind and wanted her baby back, the public was on the mother's side. The mother was the victim and Madame Restell a criminal.

Her name was also associated with the murder of Mary Rogers, the "beautiful cigar girl" whose dead body was found in the Hudson River in 1841. The case was never solved, but it served as the inspiration for the first-ever detective novel based on real events, *The Mystery of Marie Rogêt* by Edgar Allan Poe. The murderer was never named, not even by Poe's detective, C. Auguste Dupin. According to one version, aggressively promoted by articles in the tabloids, Madame Restell was the culprit. The beautiful Mary was allegedly seen entering her mansion, where she was said to have died during an abortion, her body only later winding up in the river.

The negative popularity of the abortionist continued to grow, and she gradually became a real celebrity, so much so that her mansion on the corner of Fifth Avenue became an attraction in the New York tourist guidebooks of the time. In one of the guidebooks, she was named the "Wickedest Woman in New York," in another she was listed in the chapter entitled "Child Murder."

Madame Restell's situation took a sharp turn for the worse in 1845, when a law was passed that punished abortions at all stages of pregnancy, even before the fetus showed the first signs of life. The law punished not only those who performed abortions, but also women who sought to terminate their pregnancy. The law did little to curtail attempts to terminate pregnancies, however, it merely shifted them to the back streets.

Madame Restell successfully avoided breaking the new law for two years. Then in 1847, she was accused of performing an abortion on a young woman. It did not matter that she had refused to perform the abortion at first; the truth was that she did so in the end in exchange for a very generous payment from the father. She was convicted and sentenced to one year in prison.

After she was released from prison she attempted to change her ways, applying for United States citizenship and turning her mansion into a boardinghouse. She advertised it as a place where the ladies of New York could give birth undisturbed and in comfort. The only thing she continued from her previous life was selling pills to eliminate menstrual problems. She became a respectable woman and when her daughter, Caroline, was married, the mayor of New York himself officiated at the ceremony. Madame Restell could not escape her reputation, however. Her wealth was openly aired, including all the details of her jewelry, furs, carriages, liveried coachman, and the mansion on the corner of Fifth Avenue.

In 1873, an even stricter and more puritanical law than the one from 1845 came into force. It was proposed and advocated by Anthony Comstock, a former soldier and founder of the New York Society for the Suppression of Vice. Named after him, the puritanical Comstock Law influenced morality in the U.S. for nearly 100 years. This law made it illegal to mention in public anything to do with sex. The selling or advertising of "*any article or thing designed or intended for the prevention of conception or procuring of abortion*" was considered obscene and punishable by a fine of up to $2,000 ($60,000 in today's money) and imprisonment for up to five years. We will meet the Comstock Law again in the following stories.

It was only a matter of time before this law caught up with Madame Restell. It took a while, but in March 1878 the New York police raided the mansion on the corner of Fifth Avenue on a tip from Anthony Comstock. They found pills, pamphlets about birth control, and even some instruments complete with instructions for their use. Madame Restell would never go to trial as she committed suicide on April 1 that same year. Her maid found her that morning in the bathtub, with her throat slit.

Her suicide was so unexpected that at first nobody believed the reports of her death, even assuming it was an April Fool's prank. Questions began to arise about why she had taken her own life. Madame Restell was a rich woman whose clientele was mostly upper-class and moved in the highest political circles in the city and state of New York, and she was sure to have found some salvation from justice for her indiscretion. There was speculation that it was due to fear of the investigation, which could have uncovered the suspicious circumstances of her husband's death two years earlier. Charles Lohman died rather unexpectedly and without any outward signs of illness, leaving her a wealthy widow whose estate would be worth $15 million today. Ann Lohman, alias Madame Restell, may not have had the best reputation but she was very likely the richest abortionist in history.

Story 6.2: The revolutionary and birth control

Two people, a slim, older woman and a middle-aged man, met on a December evening in 1950 in an elegant apartment high above Park Avenue in Manhattan. At the time, the woman was over 70 years old and her once beautiful red hair had long since turned white. The man was 20 years younger, with a dark mustache and unruly gray-streaked hair. She was Margaret Sanger, one of the greatest women's rights activists, and he was Gregory Goodwin Pincus, a scientist with a high IQ but a controversial reputation.

Firstly, we will get to know "*…a great woman, a courageous and indomitable person who lived to see one of the remarkable revolutions of modern times – a revolution which her torch kindled…*" as she was once described. That revolution was the sexual revolution.

Even by today's standards, let alone by the standards of the puritanical society of the early 20th century, Margaret Sanger's personal life, including her sex life, was fairly uninhibited. She had several lovers on both sides of the Atlantic and was an advocate of free love and the right to control one's own body. She refused to be a stay-at-home mother, and instead she took on a full-time role as a women's activist. She pushed for sex education, birth control, and planned parenthood, and urged women to take full control of their own bodies and decide for themselves when and how many children to have. The things we take for granted today were unthinkable in prudish America in the early 20th century. Family planning was considered immoral and, at the time, the official distribution of information about it was illegal.

She was born Margaret Louise Higgins on September 12, 1879 in New York, the sixth child in a poor Irish stonemason's family. The Higgins family were devout Christians, and Margaret lived at a time when morality was determined by the Comstock Law, the law mentioned in the previous story. This meant that it was expected for Margaret's mother to have as many children as she could manage to carry to term. She gave birth to 11, and another seven pregnancies ended in miscarriage, so it is not surprising that she died, exhausted, at the age of 49. Those 11 hungry mouths were more than Margaret's father could afford to feed, and the family lived in constant poverty. This was not an extraordinary situation. At that time, the average family had seven children, many women died in childbirth, and every fifth child did not live past the age of five.

But let us return to Margaret. After completing elementary school, she decided to take care of her sick mother. After her mother died, Margaret, then aged 22, finished nursing school. She wanted to continue her studies, but life had other plans. She met William Sanger, a nice young architect and painter, at a dance. She fell in love with him, and when they married in 1902, she left school. But Margaret was not a woman made for the classic marriage; she once declared that marriage was akin to suicide. Although they had three children, after eight years of marriage they separated and were divorced in 1921. That same year, she founded her first official organization: the American Birth Control League.

As a visiting nurse, she cared for the women living in the slums of New York's Lower East Side, the majority of whom were the families of Italian and Jewish immigrants. Their lives were similar to the life she had left behind when she left home. Every day, she encountered women who were worn out from carrying and giving birth to children and men who were unable to give all their offspring a dignified standard of living. The families lived in poverty and illness. Information about natural birth control was extremely limited, because the Comstock Law banned the dissemination of information about any form of family planning. This law was not unique in its wording, and in fact only upheld the official opinions of many. Then-president Theodore Roosevelt even declared that the use of birth control was "*criminal*" and contributed to "*race suicide.*" We already know that Margaret intended to change all this – and more.

She started off writing columns entitled "What Every Mother Should Know" and "What Every Girl Should know" for a monthly magazine. Her opinions were very frank and it is no surprise that she soon became the target of censorship by Anthony Comstock himself. Margaret Sanger was defiant, and in 1914 she established her own magazine, *The Woman Rebel*. The aim of the magazine was to provide women with all available information on birth control. Her opponents responded as expected. The federal government first ordered Margaret to stop publishing the magazine and when she ignored that order, the government confiscated three monthly issues and she was arrested and charged with promoting obscenity and violence in August 1914. If found guilty, she faced up to 45 years in prison. At the time, she was 34 years old – young, slim, red-headed and beautiful. Some compared her to Sandro Botticelli's painting of the beautiful Judith. The thought that she would have to spend the rest of her life behind bars was so frightening that she skipped bail, left her family, and fled to Europe.

Her sojourn in Europe lasted just over a year, but it was long enough to change her life completely. At the time, Europe was far less puritanical than the United States, and not only was family planning not considered immoral or illegal, it even had philosophical grounds, for instance in the work of the philosopher Thomas Malthus. Information about contraception was readily available, and Margaret soon learned more than she had learned in her entire previous life. She also fell in love with 55-year-old Henry Havelock Ellis.

Ellis was a tall, slender, elegant, and distinguished man with gray hair and a long, wavy silver beard. He was one of the world's foremost sexual psychologists and the author of the first English medical textbook on homosexuality. In fact, the term *homosexual* is attributed to him. His opinions assaulted Victorian morals and American prudishness. He approved of masturbation, and to him a woman's desire for sex and marriage was not limited just to procreation. His personal life was rather unconventional. He was a virgin when he married Edith Mary Oldham Lees, a lesbian, at the age of 32. Theirs was an open marriage. Ellis was a noted London intellectual and his friends included the philosopher Bertrand Russell and the writers George Bernard Shaw and H. G. Wells. Wells became one of Margaret's lovers and several of his stories were inspired by his relationship with the young American woman.

Ellis was not only Margaret's lover, he was also a brilliant mentor whose intellectual approach moderated her anger and hostility and taught her to think strategically. Prior to meeting Ellis, she had only an unstructured endeavor to improve the lives of women, but afterward she began to develop a concrete plan. Her primary aim was to help women obtain the right to decide for themselves how big a family they wanted, how many children they wanted to have. She believed that if sex was distinguished from having babies, it would give women the freedom to make decisions about their own lives, a freedom they could not have imagined.

When she returned to the U.S. in October 1915, one of the first things she did was to work on a suitable phrase by which she could clearly define her endeavor and her aim of distinguishing sex from reproduction. After several attempts, she decided to use the easy to understand yet benign phrase, *birth control*. On October 16, 1916, she opened the first birth control clinic in the United States in Brooklyn. She and her sister Ethel, along with several assistants, dispensed information to women about the need for, and the possibilities of, family planning, along with contraband condoms and pessaries. The clinic remained open for exactly ten days. Even though Anthony Comstock had already passed away, his law was still in force and although the charges against Margaret from 1914 had since been dropped, she did not avoid jail this time and was sentenced to 30 days for illegal distribution of birth control information and products. When they tried to fingerprint her in prison, she fought so hard that two guards allegedly had to be treated in the infirmary.

Nobody really likes prison, but Margaret and her movement were given a huge boost by her incarceration. She became a well-known and sought-after person and was invited to lecture throughout the United States. Success began to follow. In 1917, she founded *The Birth Control Review*, the first scientific journal that dealt with birth control. In 1918, her efforts were successful in allowing doctors to prescribe birth control to women in indicated cases. In 1921, she founded the American Birth Control League (and divorced William Sanger), and two years later she established the Clinical Research Bureau (CRB). This was the name under which the first official birth control clinic on the American continent operated. It was later renamed as the Birth Control Clinical Research Bureau.

During this period, she also met widower and wealthy industrialist James Noah Henry Slee. They were married after a brief engagement, but in classic Margaret fashion, she had her own conditions. The prenuptial agreement provided that the couple would live in separate apartments at separate addresses, that she would keep the name she was known by, that they would not have keys to one another's homes, and that all visits would be announced in advance. In spite of all this, Slee admired and always supported her. When she published her book *Motherhood in Bondage* in 1928, it was such a poor seller that he bought all the copies.

More important for Margaret, though, was the fact that just after their wedding she persuaded her husband to start the *Holland-Rantos Company*. This was the first company on the American continent to distribute contraceptive devices. At the time the company was established, it was illegal in the U.S. to manufacture diaphragms, so Slee smuggled them into the U.S. from Europe via Canada.

Margaret Sanger achieved possibly her greatest victory in 1936. Her supporters were growing in number and the social climate was starting to change, so she decided to take a chance. In order to provoke a decisive battle, she openly imported diaphragms from Japan. The authorities confiscated the consignment, and in 1936 Margaret once again faced charges under the Comstock Law. This time, however, the court ruled in her favor. The court repealed the Comstock Law, legalizing contraception in the U.S. Following this decision, the American Medical Association approved birth control as a part of standard medical practice in 1937 and even incorporated it into medical school textbooks.

The *Holland-Rantos Company* no longer had to smuggle contraceptives into the country and could now manufacture them. The company still exists under the slightly modified name *HR Pharmaceuticals* and it still manufactures lubricating jelly.

Margaret Sanger kept gaining allies. However, one of those relationships was so controversial that her association with the movement caused her to fall in the eyes of many people: the eugenics movement.

Eugenics is the philosophical opinion that aims to improve the genetic quality of the human population, either by increasing sexual reproduction in people with the desired qualities, or suppressing reproduction in people who do not have such qualities. That Margaret Sanger became involved in eugenics is actually understandable. She saw how poor families sank deeper into poverty with the birth of every additional child, a birth they had no control over. She believed their situation could be improved through education about how to keep the size of their families down to a number that would allow them to have a better quality of life. Because she was focused primarily on large, poor families, it may have been perceived as an effort to suppress the reproduction of this group of people. This pushed her closer and closer to a eugenics position, and there was a time when she really did believe that eugenics could greatly contribute to humanity. When her opponents also took into account her speech to a women's auxiliary of the Ku Klux Klan, her association with eugenics, and a racially oriented one at that, was solidified. Although she later backed out of the alliance, her reputation as a "notable racist and eugenicist" still holds today. When Hillary Clinton received the Margaret Sanger Award in 2009, she was sharply criticized for accepting it because was taken as a public admission of her support of eugenics.

After the victory in 1936, Margaret Sanger withdrew from the front line. Although she became the president of the newly-formed American Birth Control Federation and later helped to establish the International Planned Parenthood Federation, these were honorary rather than executive titles. She settled in Tucson,

Arizona, but she never abandoned her vision of giving women a contraceptive that would make it easy to control their sex life.

And so she met with Gregory Goodwin Pincus one December evening in 1950 in that apartment high above Park Avenue in Manhattan.

Story 6.3: The controversial biologist and his controversial experiments

The two people who met on that December evening in 1950 had a fairly long conversation. We know the woman was Margaret Sanger and that she was one of the greatest crusaders for the right of women to decide their own fate for themselves. She believed the way to get there was with birth control and family planning, but none of that was possible without a simple form of contraception. It had to be foolproof and it had to be controlled by the woman. Here to help was Gregory Pincus, the scientist with the high IQ and the shoddy reputation.

Margaret Sanger had one question for Gregory Pincus: Is it possible to create a simple and reliable contraceptive for women? He thought for a moment, then nodded and said: "*I think so. It will require a good deal of research, but it is possible.*" Sanger replied, "*Well then, start right away.*"

The work of Gregory Goodwin Pincus, or Goody as his friends called him, was relatively well-known by then but it was also controversial. His name appeared in scientific journals as well as in the daily newspapers. To his supporters, he was a well-respected scientist, but his opponents compared him to Frankenstein. Goody Pincus's line of work was in vitro fertilization. In 1934, he was successful in the in vitro fertilization of a rabbit in a test tube. Two years later, he had furthered his achievements, claiming to have begun the reproductive process in an egg without fertilization by male sperm, just by changing the environment around the egg. His achievements were provocative, as was his bearing. He soon became an inconvenience for his employer.

There had been no indications in his life or at the start of his career that things would develop in this direction. Pincus was born in 1903 in a Jewish commune in New Jersey, as the oldest of six children to farmer and teacher Joseph Pincus. As a young man, Goody imagined himself as a poet, philosopher, and a lover of women. He went to Cornell University to study agriculture, and although he was not an exemplary student, he earned a bachelor's degree. He then went to Harvard University to study biology, where he attained his highest degrees. He spent some time in Europe, in Cambridge and Berlin, then returned to Harvard where he later became an assistant professor in 1931. In 1930, he worked in Harvard's physiology lab, concentrating on mammalian reproduction. He started with rabbits. Pincus was interested in how hormones affected their reproductive process, studying fertilization and egg development. By 1934, he was able to simulate the entire reproduction process and early stages of embryo development in a test tube, outside the mother's body. He then transplanted the test tube embryo into the body of the rabbit mother, who carried the baby rabbit to term and gave birth. It was a groundbreaking discovery, and Gregory Pincus proudly presented it to the public. Two extensive articles soon appeared in the daily newspapers, with the headlines "*Rabbits Born in Glass*" and "*Bottles are Mothers*." In the second article, Pincus declared that he would repeat on humans the success he had achieved with the rabbit eggs. Goody was proud of his achievements and flattered that he was considered as some sort of "biological Edison."

A year later, he published another, even bigger discovery: he no longer needed a mother or father to develop eggs. He and his colleagues began to develop eggs in laboratory conditions without fertilization, just by manipulating the environment around the egg. This form of asexual reproduction, scientifically called parthenogenesis, is not unique in nature but it was a significant achievement because it was in a mammal. Unsurprisingly, the newspapers printed unsettling articles that can be summed up nicely with this headline: "*Manless World*." Gregory Pincus faced the uneasy reactions of the public as well as those of his colleagues. When other scientists were unable to repeat his parthenogenesis experiments, they became known by the derogatory name "Pincogenesis" and Goody Pincus became problematic. Harvard terminated his contract and nobody else was willing to employ him.

Just when he was about to give up hope, his former classmate from Harvard offered him a helping hand. He was willing to take Pincus on in his small team in the physiology department at the "rural" Clark University in Worcester, Massachusetts. His salary and research budget would both be substantially lower than what he had at Harvard and his laboratory was in a basement near a coal bin. The budget was so low that he could not even afford labels for his chemical bottles, identifying the contents by smell.

His situation was critical, with no money for research or a livelihood. But Gregory "Goody" Pincus was not one to give up easily, and when the situation was at its worst, he and his colleague elected to do something that was unprecedented in the scientific community at the time. They decided to set up their own scientific research institute. They went from door to door in Worcester, persuading people to sponsor the institute. They were great salesmen, and in 1944 they established the Worcester Foundation for Experimental Biology, investing all the money they had obtained into research. The foundation was immediately successful and attracted a growing number of young scientists. When Margaret Sanger asked him in 1950 whether it was possible to develop a contraceptive for women, he was able to reply in the affirmative.

The principle of contraception on which Gregory Pincus and his science colleagues worked was simply the application of natural laws long ago put into effect by Mother Nature herself. Beginning in puberty, a woman's ovaries produce one egg about every 28 days. If a woman has sex during this time, one of somewhere around 100 million sperm can fertilize the egg, which then attaches itself to the uterine wall where it is nourished through the woman's blood. That begins the pregnancy. The entire process is regulated by two sex hormones: estrogen and progesterone. Pincus concentrated on progesterone, which is also referred to as the pregnancy hormone. One of its roles is to prevent the ovaries from producing eggs during pregnancy, thus preventing ovulation. This makes it an effective contraceptive. So what if progesterone could be given in tablet form? It would imitate pregnancy and prevent ovulation. If no egg is released, no egg can be fertilized, meaning no pregnancy.

The action of progesterone in preventing ovulation in rabbits was first described around the time that Pincus was moving to Worcester, but nobody had looked into the practical use of the experiment. No one was interested in contraception innovations at the time. No one, that is, except Margaret Sanger and Gregory "Goody" Pincus.

Pincus and his colleagues began experimenting with progesterone on rabbits. They administered it by injection and then dissected the laboratory animals to determine whether eggs had been released from the ovaries. They had not. When they had similar results with other laboratory animals, they knew they were on the right path. In 1952, they wrote in their annual report that even progesterone administered in tablet form prevented ovulation in rabbits, with a success rate of over 90 percent. The results were so persuasive that clinical trials could begin – administering progesterone to women.

It was no easy task, as clinical trials are far more complicated than working with lab animals. The rules in the 1950s concerning clinical trials were by no means as strict as they are today, but they were still unable to find anyone willing to invest in their study. Additionally, no one wanted to support something that at the time made little sense: giving healthy women a drug just to make their lives better. However, it did make sense to American philanthropist Katharine McCormick, and after just one meeting with Pincus, on June 8, 1953, she became the chief sponsor of the entire research project. She invested the equivalent of $23 million in today's money into contraceptive tablet research. It is hard to imagine how it would have all turned out without this financial injection.

However, money was not the only problem. Gregory Pincus was a biologist, not a doctor, and as such he could do laboratory experiments but not clinical experiments. A doctor was needed, but where would they find one willing to join a controversial project run by an even more controversial scientist? It would also have to be an experienced gynecologist, someone who was willing and able to persuade women to participate in the trial and a recognized expert in the field who could bring in money for the project from pharmaceutical companies. Additionally, it could not be someone Jewish because Pincus was afraid that it would draw unwanted attention if another Jew besides himself was a part of the research.

He found Dr. John Rock, a solid-looking 62-year-old professor of gynecology at Harvard Medical School, and the first person to fertilize a human egg in a test tube. He was a Catholic, whose participation in contraception research would later earn him more than his fair share of attacks from others of that faith. Rock was a remarkable person who certainly deserves his own story in this book.

Professor John Rock was also one of few physicians with practical experience of administering progesterone to women, though he was attempting to do the exact opposite of what Gregory Pincus wanted. Rock wanted the hormone to help infertile women become pregnant. His experiment was unsuccessful, but he did prove that progesterone does not cause any significant side effects, which was a very important fact in the contraceptive project.

In early 1953, one of the boldest and most controversial clinical studies in the history of modern medicine got underway. To begin with, any form of contraception, let alone attempts to develop one, was still illegal under Massachusetts law. Both scientists faced five years imprisonment and fines of up to $1,000 if they broke the law, but Gregory Pincus came up with a way to get around it. The development of a contraceptive was officially incorporated into John Rock's project to treat infertility in women. For the same reason, the women who volunteered for the study could not be told about the study's goals. Officially, they were enrolled in a drug study to regulate their menstrual cycle – which was essentially the truth.

About 60 women were enrolled in the trial in the first year. They were women from John Rock's clinic, nurses from the clinic next door to Pincus's laboratory, and the patients of "friendly" gynecologists. The first results were disappointing, with 15 percent of the participants becoming pregnant during the study. A contraceptive that was only 85 percent effective was not exactly a breakthrough.

The main problem was that human trials are much more demanding than working with laboratory animals. The animals were kept in cages the entire time and constantly monitored, and their fate was to participate in the research planned for them by Gregory Pincus. People, on the other hand, move about freely. They often change locations, forget to take the drug, or simply stop taking it when they feel sick. Not to mention that testing a drug that manipulated the menstrual cycle was quite different from testing a drug intended to treat sick people. Sick people want a cure, they want to get better and they are willing to sacrifice their time and to deal with mild side effects. But these were healthy young women, and besides, participation was not just about taking a pill. The participants had to collect urine samples for laboratory testing, take their temperature, and do other things that interfered with their daily lives. After a year of hard work, they had a complete set of results from fewer than 30 women. It was simply not enough.

It was clear to Gregory Pincus that if he wanted his research to succeed, he had to find volunteers in places where much better cooperation could be expected. We already know that Goody was resourceful and he decided that the next round of testing would be conducted on the poor women of Puerto Rico and on psychiatric patients at the Worcester State Hospital. This did not pose a problem, because he was conducting his research at a time when strict regulations governing clinical trials were a thing of the very distant future. There were no ethics committees, and the trial subjects did not have to sign an informed consent form to participate.

It was early 1955, and his study would soon finally have enough participants. In addition, two synthetic forms of progesterone had recently been synthesized and they were substantially more effective than the natural pregnancy hormone. Everything looked more than promising, but soon summer came and he still had nothing to show for his efforts. In Puerto Rico, only ten of the original 23 participants remained and the data from those who remained were incomplete. The results from the psychiatric patients were even more distressing. John Rock had successfully tested the new synthetic progesterone in the early fall, but on only four women. It was all painfully insufficient.

The break came a year later and it required only one thing: honesty. They decided to make the goals of their study public. Puerto Rico was only a U.S. territory and not a full-fledged state, so it was not subject to the strict laws against contraceptives. They suddenly discovered that many young Puerto Rican women were more than happy about the possibility of an effective form of birth control, and they had no problem participating in a clinical study to help discover it. The study was even officially sanctioned by the government of Puerto Rico, but on the condition that there would be no publicity.

In March 1956, there were 100 Puerto Rican women enrolled in the study, increasing to 221 by the end of the year. They had all been informed of the goals of the study. Although the study was no less complicated than previous ones, the participants cooperated much more willingly than the women in the earlier trials.

The day finally came when Gregory Pincus and John Rock decided to make their achievements known officially, by going to the annual hormone research conference in September 1956. They did not choose this conference by accident. Every year, the foremost experts in endocrinology gathered at the conference, and Pincus himself had established the tradition of these scientific gatherings 13 years earlier, also becoming the chairman. Both scientists knew that if this gathering accepted their findings, they had it made.

It was John Rock who presented the findings. As a speaker, he was the complete opposite of Gregory Pincus. He was calm and held strictly to the tedious scientific facts. The word "contraception" did not come up once in his speech. However, when he finished his presentation, it could not have been clearer to the conference attendees that they had just learned of the existence of a safe and simple method for preventing ovulation, a way to allow women to prevent pregnancy.

It would take several more years for the results of the research to be put into practice, for a pill to be made available to women to prevent pregnancy. Although it had been scientifically proven, from a practical perspective contraception was still an insurmountable goal. In 1958, there were still laws in 17 U.S. states that prohibited the sale, distribution, and promotion of contraceptives. The only way around it was a disease or disorder that would justify prescribing the drug and thus, on June 10, 1957, the U.S. Food and Drug Administration (FDA) approved the sale of the pill for the treatment of menstrual irregularities. The actual birth of the contraceptive pill is May 9, 1960, when the FDA made the official announcement of approval. The next day, the *New York Times* ran a headline that circled the globe: "*U.S. Approves Pill For Birth Control*."

Story 6.4: The Catholic gynecologist and his futile hope

Christianity has strictly banned all forms of contraception since its very beginnings. The words in the first chapter of the Book of Genesis, *Be fruitful and multiply*, inextricably associated the sex act with reproduction and continues to do so to this day. If anyone dared to disobey, they would be punished by God. A case in point is the story of Onan in the Old Testament, although it tends to be misquoted.

In order to fulfill his duty under religious law, following the death of his older brother Er, Onan was to marry the brother's widow Tamar. He would only be a sort of stand-in, and any offspring would be officially deemed that of the older brother. In protest, whenever he had sex with Tamar, he "*spilled his seed on the ground*," effectively using a form of contraception that would later become known as coitus interruptus. God found that to be a sin and slew him. References to the name Onan were, wrongly, often thought to be references to masturbation (*onanisma*).

The New Testament added the precept of virginity and the institution of marriage to the duty to procreate, and therefore a young woman could "*be fruitful and multiply*" only after receiving the sacrament of marriage. This also sealed her fate: to give birth to children and care for them. Any attempt at preventing this God-given duty was considered a sin and birth control did not make an appearance in the vocabulary of the Catholic Church until the first half of the 20th century.

While the teachings of the Catholic Church were clear about birth control and remained unchanged for centuries, the life circumstances of the ordinary faithful often forced them to circumvent those teachings. At first, the main reason was health problems in women. Following numerous pregnancies, women became so worn out and weakened that they were at risk of death with another pregnancy. Social reasons came later as couples began to realize that the more children they had, the less likely it was that they would be able to care for them responsibly.

The most commonly used method of birth control in history is coitus interruptus, as mentioned in the story of Onan. The main disadvantage is that it requires the partner to take responsibility. Women had been using pessaries and diaphragms made of various materials long before they were officially invented by Wilhelm Peter Johann Mensinga of Flensburg in the 1870s. These had the advantage that family planning was controlled by the woman, so that she did not have to rely on her not-always-reliable male partner.

The oldest known condoms were discovered in the excavation of an ancient toilet at Dudley Castle near Birmingham, England. They can be dated quite precisely, since the castle burned down in 1646. In the 1950s, a box of condoms manufactured in 1780–1810 was discovered; the size of those condoms was interesting, at 7.5 inches long with a diameter of 2.4 inches. In comparison, today's European Union standards are over three quarters of an inch shorter. Remember, men were nearly eight inches shorter in the late 18th and early 19th century than they are today.

This was roughly the situation with birth control when the star of our story, John Rock, was born. He was a man of contradictions who firmly stood by his convictions, despite the fact that it not only frequently put his reputation as a doctor at risk, but also his reputation as a devout Christian. He defended himself with the advice given to him by his priest when he was 14 years old: "*John, always stick to your conscience. Never let anyone else keep it for you, and I mean anyone else.*"

John Rock was born on March 24, 1890, in the Irish community of Marlborough, Massachusetts, where he was also christened a few days later at the local Immaculate Conception Church. He was a devout Christian all his life, regularly attending mass every morning between 7:00 and 8:00. A large cross was hung over his desk. Despite this, he was called a "moral rapist" and one of his colleagues later officially requested Boston's cardinal to have Professor John Rock excommunicated.

Rock spent his childhood in the company of his sisters rather than running around and playing games with other children his own age. When he was 17, he fell madly in love with his friend, the captain of the school basketball team, but eventually he got over it and in 1925, he married. He always remained faithful to his wife, and they had five children and 19 grandchildren. If his youthful experience left any mark on him, it was an open-mindedness about various forms of sexual behavior.

With his appearance and manners, John Rock was the model of conservatism. He always dressed in a suit and starched shirt, and even on the hottest days he never left the house without a tie. He was tall, slim and silver-haired, a family doctor right out of central casting. He always held the door open for his patients and he had the same nurse for 20 years, though he never called her anything other than "Mrs. Baxter." Early in his career, he had even taken a stance against women being allowed to study medicine at Harvard University. However, neither his faith nor his conservatism could hamper his strong belief in the need for birth control. He was the only Catholic of the 15 Boston physicians who signed a petition in 1931 calling for the repeal of the ban on the sale of contraceptives.

His father owned a liquor store and had a real estate business, and he wanted his son to follow in his footsteps. John did not fulfill his father's dream, however, although he did attend a business high school and found work as an accountant. He worked briefly for a banana grower in Guatemala and then for an engineering company in Rhode Island, but was fired from both jobs and realized that he would never be a businessman. He decided to go to medical school instead, graduating from Harvard in 1918 with a specialization in obstetrics and gynecology. His decision ultimately affected the lives of thousands of women. From the start of his career as an obstetrician and gynecologist, he encountered two big, yet contradictory, wishes. His patients included women who could not get pregnant but very much wanted to, as well as women, mostly from poverty-stricken neighborhoods, who already had children and did not want any more pregnancies. He helped both groups.

For the first group of women, he established the Fertility and Endocrine Clinic at the Free Hospital for Women in Brookline, Massachusetts. It was one of the best women's fertility clinics. Rock always began by taking a detailed medical history and performing a complete medical exam. He looked at menstrual irregularities, trying to find a way to help the woman get pregnant. Where natural fertilization did not work, he used the artificial method. He was the first to fertilize a human egg in a test tube. He also understood that many cases of supposed infertility in women were, in fact, caused by the infertility of men and was one of the first to freeze and then thaw sperm without harming their potency.

During this time, he also began administering hormones to women. Rock had noticed that when a woman with fertility problems finally became pregnant, the next pregnancy was far easier. It was as though the first pregnancy had somehow balanced the hormone system. He attempted to induce a simulated pregnancy in women by injecting his patients with various combinations of progesterone and estrogen. Women's sex hormones play the key role in fertilization and pregnancy, and he wanted to administer the hormones to induce a "pseudo-pregnancy." John Rock had absolutely no experience with administering hormones, unlike Gregory Pincus from the previous story, and had never even tested them on laboratory animals. However, he first tried out every combination of hormones on himself, because he wanted to find out if they caused any pain and to make sure they were not deadly.

The good news was that after the treatment was finished, all the patients were still alive. Even better, 13 patients were later able to become pregnant. What is important for our story is that the knowledge gained by John Rock from giving hormones to women opened the door for him into Gregory Pincus's research team. It allowed him to transfer the results of the laboratory tests to clinical trials and it would link the life of John Rock closely to the contraceptive pill.

At the start of the story, we mentioned that the official Christian doctrine long kept issues of birth control at bay. Until 1930, that is. In that year, Pope Pius XI issued an encyclical called *Castii Connubii*. The Church's official opinion on any sort of prevention of procreation or birth control was clearly laid out in the encyclical. The pope called it a "*shameful and intrinsically vicious*" attempt that is "*against the law of God and of nature*," and that "*those who indulge in such are branded with the guilt of a grave sin*." The pope did, however, insert one important exception in the encyclical, saying: "*Nor are those considered as acting against nature who…use their right in the proper manner although on account of natural reasons either of time or of certain defects, new life cannot be brought forth*." His words can be – and were – interpreted to mean that a married couple can have sex for a purpose other than to make babies if the couple knew that natural reasons were preventing conception. Essentially, the Church was allowing the use of natural family planning: the prevention of pregnancy based on monitoring the signs of the fertile and infertile phases of the menstrual cycle. This is known as the rhythm method. Ninety years after the encyclical was issued, it is still the only method of contraception, apart from complete abstinence, that the Catholic Church officially recognizes as being moral.

John Rock, gynecologist and Catholic, was only being himself when he got down to putting the new method to use as soon as it was approved by the Catholic Church. He continued to treat women who wanted to have children, but now also treated women who did not. He helped them to determine precisely the intervals of their menstrual cycle and narrow down the days that were safe if they did not want to get pregnant. His clinic was the first to address the method of fertile and infertile days systematically.

John Rock became involved in the issue of birth control alongside his treatment of infertility. He was not merely a practical propagator; as a professor of gynecology at Harvard University Medical School, he added the subject of birth control to his curriculum, something unprecedented at the time. In 1949, he coauthored the book *Voluntary Parenthood*, which provided information about birth control methods for the general reader. Women who wanted to have children and women who did not want to have children passed by one another in the same hallway of his clinic.

It is not surprising that in 1952, Gregory Pincus invited him to join his team in developing hormonal contraception, the birth control pill. Rock did not have to accept the invitation. At 62, he was just about ready to retire and he had also suffered a heart attack eight years earlier. His eldest son had died in a car accident, and his beloved wife was seriously ill. But mainly he was a Catholic and taking part in this type of research could jeopardize not only his reputation, but also his personal fate. He accepted anyway.

One of the boldest, and most controversial, clinical trials in the history of modern medicine could now begin. We know from the previous story that not everything went according to the organizers' plan. We also know that it was John Rock who first informed the professional sector of the existence of hormonal contraception.

As a physician, Rock was genuinely convinced of the benefits of birth control and he championed it practically until his death. He was the one who persuaded the FDA to approve it, and even afterwards he kept on advocating for it. He appeared on television and his interviews were published in the most widely-read newspapers and magazines in the world. He traveled the country speaking and persuading, and he defended the safety of "the pill" before the United States Senate. He wrote a highly confrontational book, the title of which says it all: *The Time Has Come: A Catholic Doctor's Proposals to End the Battle over Birth Control.*

As a Catholic, he believed that hormonal contraception was essentially a variant of the rhythm method. The method of fertile and infertile days limited sexual intercourse to the safe period that existed thanks to progesterone, and the pill just prolonged the period to an entire month. He claimed that oral contraception was in effect an infertility period created by a pill. He believed it was only a matter of time before it would be accepted and approved by official church doctrine.

It seemed he would actually live to see that happen. He did not. Proponents of birth control, including John Rock, were disappointed when Pope Paul VI issued the encyclical *Humanae Vitae* on July 29, 1968. It stated that: "*...excluded is any action which either before, at the moment of, or after sexual intercourse, is specifically intended to prevent procreation – whether as an end or as a means.*"

Shortly thereafter, John Rock closed his practice, sold his house in Boston and settled on a farm in New Hampshire. He swam, sipped martinis, and listed to his favorite composers. He died at the age of 94, his life in order.

Long before his death, just after the birth control pill received approval, one of his opponents wrote to him: "*You should be afraid to meet your Maker.*" John Rock replied: "*My dear madam, in my faith, we are taught that the Lord is with us always. When my time comes, there will be no need for introductions.*"

Story 6.5: The three brilliant chemists

The Octane Number and the Black Head

In the spring of 1942, a *gringo* boarded a bus going from Mexico City to Veracruz, a city on the coast of the Gulf of Mexico. When the bus arrived at Orizaba six hours later, he transferred to another, equally ancient and rickety bus going to Cordoba. He watched the countryside going by outside the window. When he saw the road go by a small stream in a ravine, he jumped up and shouted at the driver to stop. There he found a small store. The *gringo* did not speak a word of Spanish, and the store owner, Alberto Moreno, spoke only Spanish. The conversation was brief. *Cabeza de negro?* asked the American; Alberto nodded. *Mañana*, he said. That was a word the *gringo* understood. The next morning, when he returned to Alberto's store, there were two black heads – *Cabeza de negro* – waiting for him. The American packed up the nearly 100-kilo loot, tossed the bags onto the roof of a bus and started back. Unfortunately, he did not have any permits and soon the police had confiscated his bags. But he really wanted those heads, so he bribed the police and one of the bags, weighing about 40 kilograms, was returned to him. The purpose for his trip to Mexico had been fulfilled and he could return home to Penn State in the U.S.

What was this American looking for in Mexico? What were those giant heads? No, this was not about cannibalism. *Cabeza de negro* is the name given by Mexican natives to a plant with the scientific name *Dioscorea mexicana.* It is a tuber, a distant relative of the yam, also known as *barbasco*. The woody below-ground part of the plant can grow to a diameter of three feet and sticks up about eight inches above ground. The outer layer is a thickened, wrinkled bark that from a distance looks a little like an old black man – hence the name *Cabeza de negro.*

Dioscorea mexicana, barbasco, Cabeza de negro, or black head, does not appear in this story about birth control by accident. It appears here because this exotic plant has a high content of a substance called diosgenin, which is used for synthesizing progesterone, the pregnancy hormone. Diosgenin and related substances can be found in other plants, but none contains such a high concentration as *Dioscorea*. So high, that one American felt it worthwhile to make a trek to the Mexican countryside. Without *Cabeza de negro* there would not be enough diosgenin, without diosgenin there would be no synthetic progesterone or its more effective analogues, and without those, there would be no pill.

The Pill, about which Margaret Sanger dreamed, and for which Katharine McCormick provided the financing to develop, which Gregory Pincus developed, and which John Rock spent his productive years defending. Every pill needs to have an active ingredient, and for this one, the credit goes to Russell Earl Marker, the gringo who got into that rickety bus in Mexico City on that spring day in 1942.

Russell Earl Marker was an adventurer and that bus trip to Cordoba was not the first he had taken during that time. He had arrived in Mexico a few days earlier, carrying a book on botany which told of the *Dioscorea mexicana* that was found near a stream that crossed the road between Oribaza and Cordoba. He hired a couple of Mexican botanists with a truck and together they went to find the *cabeza de negro*. At the time, however, there was widespread anti-American sentiment in Mexico and the U.S. Embassy had advised American citizens not to travel there. Russell's Mexican assistants held the same view, and fearing for their lives, they soon abandoned him and returned to the capital. It was all left up to Russell. He did not speak Spanish and did not know the local customs or rules, but he was an adventurer who tried anyway and succeeded. He was also an excellent chemist.

Marker was born in a log cabin on a poor farm near Hagerstown, Maryland, on March 12, 1902. He studied chemistry, but never completed his advanced degrees because he thought that the courses he lacked were not important for his future work. He found employment with the Ethyl Corporation in Richmond, Virginia, a company that manufactured additives for gasoline. He only stayed there for two years, but it was long enough for him to make history. He invented a way to improve gasoline quality, but mainly he developed the octane number system which is still used today to rate the quality of gasoline. It is well-known by all motorists, whether they are filling up with 92-octane "regular" at the pump in the U.S. or 95-octane "super" in Europe.

After two years, he tired of gasoline and went to work as a chemist at the Rockefeller Institute, where he spent nearly eight years researching the structure of organic compounds. After a time, he grew weary of this too and turned his attention to another area of organic chemistry, hormone research. When the head of his department at the Rockefeller Institute would not support him in this, he left the Institute and went to Penn State, taking more than a 50 percent pay cut in the process. It was nearing the end of the 1930s, a decade some chemists had named the decade of the steroid hormone. It was a time when the structures of the sex hormones, including progesterone, were being identified and a time when they could be synthetically produced. It was Russell Marker, by then a professor, who developed a chemical reaction that made it possible to produce a variety of sex hormones from the substances found in various plants. Any industrial use, however, would require a plant with a high content of those substances. He kept looking until he found it: *Dioscorea mexicana*.

Russell Marker now had a source of diosgenin and he also knew a relatively simple way of making progestin from it, not just in small laboratory quantities but in large quantities as well. He hoped he would be able to commercialize his efforts and contacted several pharmaceutical companies, but could not drum up any interest. He would have to do it himself.

He returned to Mexico and with the help of his Mexican friend, Alberto Moreno, he was able to collect and dry about ten tons of *cabeza de negro*. He found a small laboratory in Mexico City where he extracted the roots, and after evaporation he ended up with a syrup containing diosgenin in a higher concentration. He prepared the final product – synthetic progesterone – in another laboratory. He had three kilograms of it and at a price of $80 per gram, it was worth $250,000; not bad for a first attempt.

However, it was obvious that synthetic hormone production could not continue in this manner, so he began looking for Mexican entrepreneurs who might be interested in mass production. In the Mexico City telephone directory, he found a small company that worked with hormones. Its owners accepted his proposal and together they formed the company Syntex (from *Syn*thesis and M*ex*ico), with Marker taking a 40 percent shareholding in the company. Syntex would go on to become one of the largest producers of synthetic hormones on the American continent. It was acquired by pharmaceutical giant Roche in 1994.

Russell Marker was not there to see that happen. He felt he was not getting his share of the profits and when his partners would not agree to increase his share, he left the company. He also took all the know-how, even the list of the coding used on the reagent bottles. He tried to establish other companies, but after a while he seemed to have completely disappeared. When the Mexican Chemical Society presented him with an award for the development of hormone chemistry in 1969, they thought they were presenting it posthumously. Everyone was surprised when Russell Earl Marker appeared in person to accept the award. When asked what he had been doing, he replied that since 1949 he had been making and selling replicas of 18th-century silver objects.

Marker passed away nine days before his 93rd birthday, on March 3, 1995.

The Bridge Grandmaster and 106 Years of Prison

After Russell Marker slammed the door of Syntex behind him, it was the beginning of hard times for the remaining owners. The formula, and everything else they needed to make progesterone, was only inside the inventor's head and in the materials he had taken with him. They had to find a replacement, and fast.

Russell Earl Marker apparently did not have an easygoing nature. However, had he not been the way he was, the next hero of our story would never have entered the picture: George Rosenkranz, another brilliant chemist.

He was born Rosenkranz Györgyi on August 20, 1916, in Budapest, Hungary. He was educated at the Swiss Federal Institute of Technology (*Eidgenössische Technische Hochschule*) in Zurich, as one of a group led by a future Nobel laureate of Croatian descent, Lavoslav Ružička. The group had several undergraduate and doctoral students of Jewish descent, and for quite some time Ružička was able to protect them from the anti-Semitic mood that prevailed after the rise of Nazism in Europe. The longer Rosenkranz stayed in Zurich, however, the more dangerous it was for him, so he decided to emigrate. Friends arranged for a professorship for him far from the dangers of Europe, at the university in Ecuador's capital, Quito, and in late 1941 he boarded a ship headed for Cuba. There he was to board another steamer that would take him to his new home, but that never happened. The Japanese attacked Pearl Harbor, America entered the war, and the ship did not set sail for Ecuador. George Rosenkranz spent the next four years in Cuba, and while looking for work he knocked on the door of the biggest local pharmaceutical company. At first, they turned him down, saying that they had never needed a chemist in all these years, but they did not regret the decision to change their minds and hire him. Thanks to their new chemist, the company soon developed a drug to treat venereal disease. Before the introduction of penicillin, it was a real commercial hit. The company made money, and the poor young chemist became a wealthy and successful man.

It was not long before they heard about his expertise in not-so-distant Mexico, and it was only a matter of time before the owners of Syntex were able to entice the young chemist to join their company. It also helped that he got along famously with one of the owners by the name of Emeric Somlo, who had previously been a lawyer and, just like George Rosenkranz, was Hungarian. He had emigrated to Mexico in 1928 to start up a drug importing business, where he established a small pharmaceutical company – the one Russell Marker had discovered in the telephone directory.

When the young scientist took a job as a director with Syntex in 1945, no one could have known that he would remain with the company for 37 years. Rosenkranz was given a staff of nine lab assistants, three strong men, and one chemist. There was also a set of bottles marked with some unidentifiable code and he was tasked with reverse engineering Russell Marker's synthetic progesterone. As he later said, "chemical archaeology" was not his forte and so he decided to develop his own processes. He succeeded and soon they were back in business with hormone production. They did not stop at producing progesterone, however. Syntex gradually began to produce other sex hormones.

Rosenkranz could not do everything himself, and he needed a team of capable people. One of the young talented chemists who accepted his offer to join Syntex was Carl Djerassi, the third hero of this story. However, we must first finish the story of George Rosenkranz. Syntex slowly began to change from a fellowship of enthusiastic amateurs to a serious firm. Profits from the development and production of sex hormones were huge, and the company began to grow, as did its need for raw materials for mass production. The only place to find *Dioscorea* was in the wilderness of southeastern Mexico, giving rise to the massive *barbasco* trade. The giant tubers were dug up by poor *barbasqueros* (yam pickers), who would sell them to buyers, who in turn supplied them to the local manufacturers. They then produced a thick syrup from them and the syrup was crystalized into diosgenin, which was then supplied to Syntex where they finalized the entire process and prepared the synthetic hormones. Syntex created the industry of chemical hormone production in Mexico, which would grow to impressive proportions. In the 1970s, the livelihood of 125,000 peasant farmers depended on the *barbasco* trade, and those of thousands more on the processing of the yams.

George Rosenkranz left the research to his younger colleagues and became executive director. The company branched out, its expansion culminating in 1964 when it moved operations from Mexico to the Stanford Research Park in Palo Alto, California. The value of the company kept growing, and when the pharmaceutical giant Roche decided to acquire Syntex in 1994, it paid a remarkable $5.3 billion for the pleasure. When George Rosenkranz had joined the company in 1945, it was a small family firm valued at a few hundred thousand dollars.

Rosenkranz wrote over 300 research articles and held 150 patents. He also wrote 14 books, though they were not about chemistry, but about his second love, bridge.

Even after stepping down as executive director of Syntex in 1982, Rosenkranz remained active in his profession, but it was his hobby that he dedicated his time to. He was a bridge champion, and not just one of the best in Mexico, but worldwide. He won 12 American championships, and was the captain of the Mexican team in two U.S. championships. He is an honorary member and Grand Life Master in the American Contract Bridge League and was inducted into the Bridge Hall of Fame in 2000. He was also a recognized bridge theorist and invented a bidding process which is named after him: the Rosenkranz double.

He had many partners throughout his bridge career, but his favorite partner was his wife, Edith. On one occasion, they were playing bridge in the summer tournament of the North American Bridge Championships on July 19, 1984, at the Sheraton in Washington, D.C. When game day ended some time before midnight, the teams returned to their rooms, but Edith did not return to hers. Rosenkranz soon received a phone call informing him that his wife had been kidnapped and that they wanted a million dollars in ransom. Despite the risks, he decided to notify the FBI of the kidnapping and when he took the money to the agreed location, agents were waiting in the dark for the kidnappers. They were arrested and Mrs. Rosenkranz was released. The money was also recovered and the next day, George Rosenkranz continued playing in the tournament.

To everyone's astonishment, the leader of the kidnappers turned out to be a colleague who had played in the tournament. He was sentenced to 106 years in prison for kidnapping.

George Rosenkranz lived quite a long life. Even at 90, he regularly went to the gym and played the piano. He passed away two months before his 103rd birthday, on June 23, 2019.

The Father Who Would Rather Be a Mother

We have already briefly mentioned the name of the third brilliant chemist in this story, Carl Djerassi. George Rosenkranz had enticed him to join Syntex when he had been in need of capable young chemists and he definitely fitted the bill. Not only a chemist, but a self-proclaimed "intellectual polygamist." He was a novelist, playwright, poet, philanthropist, art collector, and entrepreneur. He was also one of the most published chemists in history. He was best known, however, for helping to spark the sexual and social revolution and being the first to synthesize the active ingredient in the oral contraceptive.

But first things first. Carl Djerassi was born on October 29, 1923, in Vienna, Austria, into a family of Jewish physicians. His father specialized in venereal diseases and his mother was a dentist. His parents soon divorced and his father went back to his native Bulgaria. Carl alternated between living with his mother in Vienna and with his father in Sofia. In 1938, the situation became unsafe for the Jewish residents of Vienna, and Carl's father decided to help his ex-wife and son. He returned to Vienna and remarried Carl's mother, allowing them to move immediately to Sofia. Bulgaria was one of the few countries that protected its Jewish community from deportation to the concentration camps. The marriage was annulled two days later, but Carl and his mother were safe.

They stayed in Sofia for just over a year. The U.S. had a Jewish quota at the time, and Djerassi and his mother were lucky enough to be included in that quota, so they moved to the United States. They arrived practically without any money, and they were relieved of their last $20 by an underhanded taxi driver. Thanks to Eleanor Roosevelt, the wife of the American president, Djerassi received a scholarship that was enough for him to get a degree in chemistry. He graduated with honors, and right after he finished school he took a job with the American branch of the pharmaceutical company Ciba. Within a year, he was already involved in work on the development of Pyribenzamine, the first commercially available antihistamine. This drug revolutionized the treatment of asthma, hay fever, hives, and other allergies. It was the first of Carl Djerassi's numerous patents.

His life took on a new direction in 1949, when he accepted the invitation of George Rosenkranz to join Syntex. It was just when Rosenkranz was starting up production of progesterone and other hormones, and he needed skilled chemists to expand the company. With a few leaves of absence to work in academics, Djerassi stayed at Syntex until 1972, working tirelessly. Djerassi's older colleague once good-naturedly said about him that there were 25 hours in his day. He also had at his disposal a large amount of the diosgenin that had been discovered in *cabeza de negro* by chemist and adventurer Russell Marker.

The year 1951 witnessed some great discoveries. The first was the synthesis of the steroid hormone cortisone in June. More than one large pharmaceutical company had been racing to synthesize it but the upstart company from Mexico City won. Later that year, on October 15, the first synthetic analogue of progesterone, *norethisterone*, was prepared for the first time. *Norethisterone,* or *norethindrone,* is more stable and far more effective than progesterone, and it was one of the two synthetic analogues of progesterone that were later used as the active ingredient in oral contraceptives. Carl Djerassi was not yet 28 years old on that day but he was later nicknamed the "father of the pill" for his discovery. He would dismiss that title, saying with a smile that if he was the father, then where was the mother. He added that three people were needed to give birth: a father, a mother, and a midwife.

He was probably more the mother than the father, because he provided the egg. That would make Gregory Pincus the father, and the midwife – the person who brought the child into the world – was therefore the gynecologist John Rock, according to Carl Djerassi.

Djerassi would spend the remainder, in fact the majority, of his career alternating between professor of chemistry at Stanford University in Palo Alto and director of research at Syntex in Mexico. The extent of his contributions to research is immeasurable, as evidenced by more than 1,000 scientific papers. His participation in the discoveries made by Syntex made Carl Djerassi a rich man and allowed him to purchase 1,200 acres near Palo Alto where he started a cattle ranch he called SMIP, initially an acronym for *Syntex makes it possible*. He also invested in art and became a fervent collector. His collection of modern art was considered one of the largest private collections, and he donated a portion of his 150 Paul Klee works to the Albertina museum in Vienna.

Djerassi was married three times. He married his first wife Virginia Jeremiah when he was 19, and the marriage lasted a little less than seven years. He got a quickie divorce in Mexico because the woman he had started an affair with, Norma Lundholm, became pregnant. The pill had not yet been invented. His second marriage lasted substantially longer, at 26 years. His third marriage was to writer Diane Middlebrook. A turning point in his life came in 1978, when his daughter Pamela, a gifted young artist, took her own life. He sold his most valuable paintings and turned SMIP into an artists' colony and center of modern art.

In addition to the huge amount of scientific papers, he was also a prolific author in other areas. He wrote five novels, seven plays, a book of essays, and several poems.

During his life, Djerassi received over 20 science and literature awards, honors and numerous honorary doctorates. Austria issued a stamp in his honor, and an iceberg in the Antarctic was named after him. There was one honor he did not receive. He once said he would like to be the first 100-year-old professor of history at Stanford University, but that was not to be. He died in 2015 when he was "only" 91.

Concluding remarks

The first person to call the oral contraceptive *The Pill* was writer Aldous Huxley in one of his novels in 1958. The name was popularized and to this day it is used in place of the more verbose "oral contraceptive" or "contraceptive pill." In the years since it was invented, the pill has also become an important social and cultural phenomenon. It was a symbol of the sexual and social revolution of the 1960s.

According to the UN report on trends in the use of contraceptives of 2015, 64 percent of married or in-union women used some form of contraception. Over 100 million women of reproductive age chose the pill, with Europe and the U.S. making up about one-third of that usage. The pill was used more often than classic methods such as condoms, coitus interruptus, or the rhythm method.

One last thing about the parents of the pill. We already know that Carl Djerassi took issue with being given the position of father and that he would have rather been the mother, and have determined that he considered Gregory Pincus the father, and John Rock the midwife who brought the child into the world. But there are two godmothers we must not forget about because without them, who knows what the fate of the pill might have been. The first is Margaret Sanger, who provided the vision, and the second is Katharine McCormick, who provided the funding.

7

Chlorpromazine

To paraphrase the Wikipedia entry on the subject, psychiatry is a field of medicine concerned with the study, diagnosis, treatment and prevention of mental disorders. These include various affective or mood disorders, behavioral disorders, and cognitive disorders.

The word psychiatry is derived from two ancient Greek words: *psych(e)* meaning soul, and *iatry (iatreía)* meaning medical treatment or healing. Although the treatment of "lunatics," whatever it may have been, can be traced back to ancient civilizations, the term psychiatry first appeared only in the early 19th century. Throughout its history, it was more of a disparaged rather than a venerated medical specialization. Even as recently as the 1960s, efforts were still being made to exclude psychiatry from the list of medical fields. Fortunately that did not happen, because the way things are now, many of us are going to need it in the near future. According to the World Health Organization (WHO) and the National Institute of Mental Health (NIMH), 83 million adults in Europe and 44 million adults in America annually are already suffering from at least one mental disorder. The numbers will continue to rise instead of fall.

Chlorpromazine was the first drug to eliminate the majority of symptoms in patients with serious mental disorders. The discovery was groundbreaking and it started a new era of psychiatry in which psychiatrists could now not only help, but also heal.

Before the more-or-less accidental discovery of the benefits of chlorpromazine, there was a whole slew of experiments in the treatment of mentally ill patients throughout the history of psychiatry. Experiments that were frequently bizarre and even barbaric from today's perspective. But if these are judged according to the knowledge that was available in their day, their efforts to help patients cannot be

V. Marko, *From Aspirin to Viagra*, Springer Praxis Books,
https://doi.org/10.1007/978-3-030-44286-6_7

denied, even though they were sometimes painful and not always safe. Compared to drugs for the treatment of somatic problems, chlorpromazine arrived on the scene fairly late. At the time its medicinal properties were discovered, the effects of vaccines, quinine, aspirin, insulin, penicillin, and many other drugs had long been known and had come into common use. Perhaps the treatment of mental disorders is a bit more complicated.

Story 7.1: The enlightened doctor and freeing the insane

There is a painting on the wall near the entrance to the library of the Salpêtrière Hospital, located in the 13th arrondissement of Paris. An oil on canvas, it is composed of three parts. The figures of five women are on the right, some sitting, others lying on the trampled dirt. They are depicted in five poses, each representing one of the mental disorders. One is leaning against a post, trapped in her own melancholy; another has a glint of panic in her eyes; the third sits with a vacant look; the fourth is in the throes of a manic fit; and the fifth is tearing at her clothes in her madness. All are shackled with iron chains to posts, inside a sizable courtyard that appears to be a part of a large asylum. The entire scene is brightly lit, each detail clearly distinguishable.

The main story takes place in the center of the canvas. The central figure is a young woman being unchained by an assistant. The woman is standing passively, her hair disheveled and her dress in disarray. Her stance demonstrates her distance from what is happening, as though she were not aware.

On the left, partially concealed in the shadows, stands a group of onlookers. Some are asylum employees, others visitors to the asylum. A man stands out among them, wearing dark garments and a tricorn hat, a cane in his hand. He is clearly the one in charge of what is going on.

This oil painting measures about 5 × 3 feet. Painted by the French artist Tony Robert-Fleury, it is entitled, in French, *Pinel, médecin en chef de la Salpêtrière, délivrant les aliénés de leurs chaînes* (*Pinel, chief medical officer, releasing the lunatics from their chains*). It is better known under the title *Pinel freeing the insane*.

It was painted in 1876, about 75 years after the incident is supposed to have taken place, but it does not depict an actual event. Instead, it is a legend that portrays Philippe Pinel as a humanistic revolutionary who exposed mental illness to the sunlight. It is a legend that persists, and one that was accepted by entire generations of medical students. Philippe Pinel was one of the most notable characters in the history of psychiatry, and he does deserve to be admired and respected. He contributed significantly to a change in the attitude toward the mental patients locked behind the walls of the asylums that were built for them.

People are generally willing to accept differences in others, but when those differences exceed the level of tolerance, there is a sudden and intense need to force such a person out of society. The limits of what people are willing to accept is a very individual thing, influenced by the culture and degree of advancement of the society.

Lunatics, madmen, crackpots, nutjobs, morons, idiots, psychos – the synonyms do not end there – are an inseparable part of the human race. These are people, in today's terminology known to be suffering with mental disorders, who have always been shunned by society. While there is a certain romanticized image of madness as a form of living free with reckless abandon, or about lunatics living out in the world and the community taking care of them, the reality is far harsher. The main burden of caring for them had always fallen to the family, but life in earlier times was difficult even for those who were healthy. The fates of people with mental illnesses were much, much worse and most spent their entire lives locked in basements or barns, often tied up or chained. Other forms of care were similar.

The so-called "ships of fools" were the first mention of group institutionalization for lunatics. They were allegedly boats that roamed along the rivers, taking the mentally ill from town to town. Although it is allegory, originating from Book VI of Plato's *Republic*, taken later by medieval authors and beautifully brought to life by Hieronymus Bosch and Albrecht Dürer, it does aptly represent how the society of the day thought the mentally ill should be cared for. They were to be excluded from "normal" society and, if possible, left to their own devices. As we know, these tendencies are still prevalent.

The first land-based "ships of fools" – asylums – began to appear in the 13th century, but it was not until around the 18th century that they began to flourish. Their main purpose was the same as with home care: keep separate, lock up, and often provide just a minimum of care. Chains, straitjackets, and other forms of restraint were the norm in these institutions. Hygiene was extremely poor and asylums reeked of human sweat and feces. Madness was considered a state, not an illness, so for many years treatment was not even a consideration. The history of psychiatry began on both shores of the Atlantic as an era of guarded asylums and institutions, where individuals who were a danger to themselves – and inconvenient to others – were locked away.

The first signs of change began to appear in the second half of the 18th century, particularly under the expanding influence of the thinking of enlightenment. Gradually, slowly, asylum patients started to be viewed not as inconvenient creatures that must be isolated, but as human beings who must, and could, be helped.

It was during this time that Philippe Pinel arrived on the scene. He was born on April 25, 1745, in Jonquières in the south of France. His father was a barber surgeon, which was neither a profitable nor a respected profession. The young

Philippe at first studied theology and philosophy and did not begin to study medicine and mathematics until he was 25 years old. He attended universities in Toulouse and Montpellier, and after obtaining a license to practice medicine, he went to Paris in 1778. He was unable to practice medicine there, because the bureaucratic regulations of the centrally-governed France did not recognize medical degrees from provincial schools, so he worked outside of practical medicine for a full 15 years. Pinel wrote articles, translated scientific papers into French, and edited a medical journal. His wide-ranging interests opened the door for him to the intellectual circles of Paris. Around that time, his interest in mental disorders was becoming apparent and he began to visit and examine patients in private sanatoriums.

A personal experience influenced Pinel's selection of a specialization and his entire later life, as well as the lives of thousands of the mentally ill. A friend of his had developed a nervous melancholy that turned into a mania, causing the friend to run into the woods where he was ripped apart by wolves, at least according to one of Pinel's biographers.

The French Revolution was the turning point in Pinel's life. He was a silent proponent, filled with thoughts of enlightenment, his mind full of reform ideas. He was carried upwards on the Revolutionary wave, and he was finally given the opportunity to put his medical education to use. On August 25, 1793, Pinel was appointed physician of the infirmaries at Bicêtre Hospital in the southern part of Paris.

At the time, *L'hôpital Bicêtre* was more of an asylum than a real hospital. When Philippe Pinel was appointed it housed roughly 4,000 men, a motley assortment of criminals, small-time con artists and syphilitics, as well as around 200 mental patients. The conditions were atrocious. Patients were chained to the walls, sometimes for decades. They were cared for no better than animals and that was the prevailing attitude toward them. The residents of Paris were allowed to go and view them, much like people today go to the zoo to look at animals.

Philippe Pinel took over the mental ward and began to implement his visions, most of which are a given today, but at the time were considered revolutionary. He eschewed violent methods, improved hygiene, and implemented exercise and purposeful work. Instead of violence and judgment, he favored gentleness and understanding. He supported therapy that required close personal contact and intense observation of patients. His approach made history under the term moral treatment (*traitement moral*). Some authors prefer the term "psychological approach." They view it as one of the first attempts at psychotherapy.

After spending two years at the Bicêtre, Pinel moved to Pitié-Salpêtrière Hospital (*Groupe hospitalier Pitié-Salpêtrière*), now as chief doctor. The name of the hospital is derived from its original purpose as a gunpowder factory; at the time, gunpowder was made from saltpeter, or *salpêtre* in French. Whereas the

Bicêtre Hospital was a facility for men, his new workplace was a women's hospital, although "hospital" is not exactly the right word to describe what it was used for. At the time, Salpêtrière was a large hospice, an expansive space with many pavilions housing around 7,000 women – the insane, crippled, chronically ill, mentally retarded, and even unmarried pregnant women and old and impoverished women. Pinel continued to implement his visions in this new hospital. Among other things, he began vaccinating the patients in 1800 and this is the time period associated with the legend of how he was *"freeing the insane"* as painted about 75 years later by Tony Robert-Fleury. The truth is that there were cases where his approach was so successful that he was able, apparently as an example, to remove the chains from some patients and take them from the depths of darkness to the light of day. Those cases were few and far between, however, and restraint was a method that continued to be used on patients.

Pinel's primary contribution to psychiatry, and to the mentally ill, was not in the removal of chains from patients in the physical sense of the word, even though the legend that describes this act is wonderfully romantic. Nor was it the fact that he improved the conditions in hospices and implemented psychosocial therapeutic principles, although with that alone he would have found fame. His greatest contribution was that, more than anyone else before him, he helped change how mental illnesses were viewed. He helped turn "lunatics" into patients who needed care and understanding. He discovered that it was possible to be sensible and purposeful even in the case of patients whose behavior was difficult to understand. His contribution was that he removed the proverbial chains – symbolically.

Story 7.2: Many attempts and difficult beginnings for treatment

The late 19th century was groundbreaking in terms of the discovery of the invisible culprits behind the various brutal epidemics that had been devastating the human population for centuries. The work of Robert Koch, Louis Pasteur, Ronald Ross and numerous other scientists clarified the role of microorganisms in causing and spreading the diseases responsible for the deaths of many millions of people. That allowed others like Paul Ehrlich, Gerhard Domagk, Alexander Fleming and many more to discover ways to win, often triumphantly, the battle against deadly diseases.

But there was no such luck with mental disorders, because they are not caused by any microorganisms, bacteria, viruses, or other microscopic foes. All of those advancements had practically no effect on the treatment of mental illnesses and disorders. However, psychiatrists were also physicians who were concerned about the fate of their patients, and they were willing to try everything possible – and impossible – to help patients suffering from mental disorders. Unfortunately, they had no knowledge of the causes of the disorders, or of how the various substances worked in the human brain, so most of the discoveries were made purely by accident. This chapter recounts those accidents, most of which wound up being dead-ends.

The situation in psychiatric institutions and asylums had not changed significantly over the centuries since the revolutionary approach introduced by Philippe Pinel. The institutions were full of old, sick, and imprisoned wretches, and the number of patients with mental illness continued to rise in these bleak establishments. From 1857 to 1909, the number of inmates held in such institutions doubled in England, and each institution had an average population of over 1,000. This growth continued at the same pace in the 20th century. Between 1903 and 1933, the number of patients committed to mental hospitals in the U.S. increased from 143,000 to 366,000.

The discovery of the sedative effect of alkaloids resulted in the first, imperceptible progress in psychopharmacology. Morphine, scopolamine, and atropine, usually as part of various mixtures and cocktails, slowly found a use in the control of serious manic states. Unfortunately, alkaloids were also comparatively expensive and that made their widespread use for the thousands of impoverished residents of the institutions relatively unfeasible. A cheaper solution would have to be found and it came from the discovery of the sedative effects of two simple compounds – chloral hydrate and potassium bromide. Potassium bromide was the main actor in the first real experiment to control a mental illness. It took place in 1897 in China, and as expected, it was an accidental discovery.

The experiment was conducted by Scottish physician Neil Macleod, who had opened a medical practice in Shanghai in 1879. About 20 years later, a certain respectable English family asked a discreet favor of him. He was to bring a member of their family from Japan to Shanghai. Far from home, she had developed an acute mania. Dr. Macleod had nobody to help him accomplish this difficult task, so he decided to have her heavily sedated. She was given a large dose of potassium bromide that caused her to fall asleep, and in this condition she was able to make the long journey lasting several days from Japan to China. When she awoke, Macleod was surprised to discover that she no longer had any symptoms of mental distress. When she had a repeat episode of mania several years later, Dr. Macleod repeated the sleep therapy. After sleeping for 32 days, the woman once again awoke healthy. Despite the success, "the bromide sleep" never made much of a splash as a treatment for mania. The drug was too toxic and the entire process too risky. However, it was the first time in the history of psychiatry that a drug was used to alleviate the symptoms of a mental disorder.

The method experienced a renaissance about 15 years later, when a new class of sedatives was used to induce a deep sleep: barbiturates. Their sedative effects had been known since the early 20th century, and the first of the barbiturates, barbital, was soon commonly marketed under the names Veronal and Medinal.

Barbituric acid, the basis of all barbiturates, was first synthesized in 1864. One of the precursors is urea, which makes up the second half of the name of this acid. The first half was derived from the name Barbara. There are three different stories about who that Barbara might have been. In the first story, she was the girlfriend of the chemist who synthesized the acid. In the second story, Barbara was a waitress who provided her urine to prepare the acid. In the third, she is Saint Barbara, the patron saint of artillery officers. Coincidentally, an artillery unit was honoring their patron saint in the same pub where the new compound was being celebrated, bringing together artillery and chemistry for a brief period of time. Veronal was named for the city of Verona. According to its discoverer, this Italian city was the most hypnotic of all the cities he had ever visited.

The first use of barbiturates to induce a deep medicinal sleep was described in 1915, but it did not become widespread until the 1920s. Swiss physician Jacob Klaesi is credited with making it popular. Like Macleod 20 years earlier, Klaesi also did not initially intend to use deep sleep as a therapeutic method. He was an adherent of psychotherapy and he only wanted to use sleep to eliminate communication barriers with patients. Also like his predecessor, he was surprised by how deep sleep could alleviate and even eliminate the symptoms of serious mental disorders. At first, sleep induced by barbiturates was relatively unsafe, but as safer derivatives of barbituric acid were developed, sleep therapy became the first therapy that offered real hope for improving the condition of patients in asylums. It was used until the 1950s. Then, in the 1960s, safer and more effective methods were introduced and the use of sleep therapy for mental disorders gradually declined, until it was completely discontinued in the late 1970s.

The next type of therapy began in Berlin in the 1930s with an Austrian psychiatrist named Manfred Sakel. Known as insulin shock therapy, or insulin coma therapy, the first experiments with this treatment were successfully used to treat people with a morphine addiction. Not only did the therapy make withdrawal symptoms easier for the patients to handle, they also became less impulsive and excitable.

After moving to the university clinic in Vienna, Manfred Sakel began to experiment with insulin shock in schizophrenic patients. It was a difficult method and required an experienced medical team. Patients received insulin injections to induce a hypoglycemic coma, which they were brought out of by the administration of glucose. The coma lasted 20–60 minutes, but the most important part of the treatment was correctly judging the time to end the coma. Patients often underwent as many as 50 comas, with the course of treatment lasting up to two years. While the process may seem pretty scary, patients handled the treatment without serious complications. The method proved to be relatively successful in treating schizophrenia, even though the therapeutic effect was never satisfactorily explained.

At the time the first reports were published, insulin shock therapy was so unusual (as it would be today) that at first, Manfred Sakel was considered a quack and the results of his work were not accepted. One of his coworkers later confessed to having slept with his passport under his pillow so he could flee in case one of the patients did not survive the therapy. However, the method gained footing and soon it was being practiced throughout Europe. In the 1940s and 1950s, it crossed the Atlantic and became widespread in the U.S. as well. In the early 1960s, there were insulin units in over 100 American psychiatric hospitals. The development of new drugs in the early 1960s brought an end to this therapy too.

Just two years after the positive results of insulin shock therapy were published, another similar method appeared. It is known as convulsive therapy and was a method developed by Ladislas von Meduna (born as Meduna László), wherein he induced epileptic seizures as treatment.

This method was based on the false assumption that epilepsy and schizophrenia are mutually exclusive. While the connection between epilepsy and schizophrenia may seem unlikely on first glance, there were signs that such a connection was possible. Microscopic analysis of the brains of patients with schizophrenia and those who had had epileptic seizures suggested a link. There was a very low incidence of epilepsy in schizophrenic patients, and in some cases, schizophrenia receded when the patient began to have epileptic seizures. Could inducing epileptic seizures in schizophrenics eliminate the disease?

On the basis of that information, Ladislas von Meduna began treating schizophrenia by inducing convulsions. Camphor was the first substance he used to do so, first testing it on laboratory animals, and then on Tuesday, January 23, 1934, he injected camphor into the first patient. The convulsion began 45 seconds later and lasted another 60 seconds. The patient's condition did not improve after the first injection, so they repeated the procedure several more times over the course of the next weeks. On Saturday, February 10, the patient awoke without any signs of schizophrenia.

According to the notes from that time, after completing treatment the patient felt so good that he escaped from the institution, but upon arriving home he caught his wife with a lover. He beat them both up and said he would rather go back to the institution where there was peace and he was treated with respect.

Success with the first patient emboldened Meduna and he continued to use convulsive therapy. He used it to treat more than 100 psychiatric patients over the course of several years with a relatively high success rate of nearly 50 percent. Because there were various adverse effects with camphor, he later replaced it with pentylenetetrazol, a substance more commonly marketed under the names Metrazol and Cardiazol, which was used as a circulatory stimulant.

Despite Meduna's positive results, this therapy never caught on in Europe like the previous method had, and eventually the chemical inducement of convulsions was replaced by a new method. They began to induce convulsions with electrical charges.

Electroconvulsive therapy was launched on April 18, 1938. On that Monday morning, in an empty room on the first floor of the Clinic of Mental and Neurological Diseases at the University of Rome, a small group of people gathered: the chair of the clinic, Professor Ugo Cerletti; his assistant, Dr. Lucio Bini; and another three physicians and two attendants. One of the assistants had the very important task of guarding the door to prevent any unauthorized entry. The other six gathered around the bed of the eighth person, Enrico X, a 39-year-old technician who had been brought in a few days earlier by the police in a state of confusion, complaining of being "*telepathically controlled*." He had no idea he would soon become the first patient to be successfully cured of a mental disorder with electroshock therapy.

A device on a stand stood next to the bed, a white box with buttons and indicators. It had a short cable with two electrodes at the end, not unlike headphones. One of the attendants placed the electrodes on Enrico X's temples and Lucio Bini turned on the device. He slowly increased the voltage in short impulses lasting a tenth of a second. When the device reached 110 volts, the patient went into convulsions that lasted 48 seconds. Afterwards, the patient came around, calm and in a good mood.

That Monday in April saw the culmination of almost ten years of research on the safe application of electrical charges to induce convulsions by Ugo Cerletti. "Maestro", as his colleagues called him, began his research most untraditionally – in a slaughterhouse where pigs were butchered – and continued with many experiments on laboratory animals. He eventually found safer parts of the body on which to place the electrodes and fine-tuned the voltage needed to induce convulsions. Enrico X not only survived the therapy, his condition was significantly improved. However, Enrico X was not the first patient in history to undergo electroshock therapy. It was first described in the first century AD, and it was used to treat insomnia. The source of the electricity was the electric ray, with the fish being placed on the patient's temples. This type of fish is capable of producing a discharge of around 30 volts, something the patient would definitely have felt.

News of the success of Ugo Cerletti and his team in treating a mental disorder spread quickly. One year after the device was first used in Rome, there were similar devices in use in France and England, and by 1940 the New York State Psychiatric Institute was using it too. That same year, electroshock therapy for treating psychoses was also implemented in other countries, including the author's country, Slovakia. The method was shown to be effective not only in the treatment of schizophrenia, but also in bipolar disorder and major depressive disorder. As experience with inducing convulsions grew, so the devices used in the treatment were developed and improved. The devices used in modern times bear no resemblance to the little white box constructed in 1938 by Lucio Bini. Today, even at a time when pharmacotherapy is highly advanced, electroconvulsive therapy (ECT) is a safe, fast-acting, and very effective method of treatment for severe mental disorders. In fact, it is considered the safest treatment method for severe depression during pregnancy. An estimated one million patients are treated with ECT every year.

A little more about the multi-talented Ugo Cerletti. During World War I, he suggested the use of white winter uniforms for alpine troops instead of the dark uniforms they had been using. He also built a carillon with a tonal scale from empty artillery casings.

The final method presented in this story is the most unfortunate of all the attempts to treat the mentally ill, and that is despite the fact that the person who introduced the method, Portuguese neurologist Egaz Moniz (born António Caetano de Abreu Freire), received the Nobel Prize for it. The method is known as a lobotomy.

According to the Encyclopaedia Britannica, a lobotomy is a "*surgical procedure in which the nerve pathways in a lobe or lobes of the brain are severed from those in other areas. The procedure was formerly used as a radical therapeutic measure to help grossly disturbed patients with schizophrenia, manic depression and mania (bipolar disorder), and other mental illnesses*."

It is hard to understand today the enthusiasm with which the lobotomy was received (Moniz had initially named it leucotomy). Actually, the same applies to all the methods described earlier, but at the time they were developed, physicians were stymied in their efforts to help patients with severe mental disorders. It is not surprising that when Egaz Moniz heard at a conference about a surgical procedure that calmed down aggressive chimpanzees, he began to think about how the same procedure could be used to treat severe mental disorders in humans. He did not ponder very long. Three months later, on November 12, 1935, he performed the first leucotomy on a psychiatric patient. He injected absolute alcohol into the patient's frontal lobes through each of six holes created on either side of the skull. The intention was to create a sort of barrier that was supposed to have separated the front part of the brain from the rest. However, Moniz was not satisfied with the procedure and instead he designed a special instrument he named a leucotome. It was about the size of an injection needle and about four inches long with a retractable wire loop. The leucotome was inserted through a hole in the skull into the patient's brain and then slowly rotated to make a round cut in the brain tissue.

Egaz Moniz and his team operated on 20 patients with various psychiatric illnesses over the course of six months, and the results were promising. Seven patients were cured and seven others improved, although six showed no change.

Up to that point, it looked like there was a bright future for the prefrontal leucotomy. Things changed significantly when it was adopted enthusiastically by American physicians. They began to modify the relatively difficult operation, replacing the leucotome with something similar to an icepick, and the *lobotomy* – the American term for Moniz's leucotomy – became a routine surgery commonly performed by just one doctor. While in Europe the leucotomy was still a relatively rare procedure, the American lobotomy was performed on almost 19,000 patients between 1936 and 1951.

The method was abandoned in the early 1950s, just as quickly as it had been enthusiastically adopted. In retrospect, the main reasons that the lobotomy was indefensible were ethical. It is true that it could be used, and often successfully, to treat patients in whom other methods either could not be used or were ineffective. It is also true that it succeeded in quieting agitated patients. On the other hand, it also caused them to lose their judgment and their social habits, and it increased irritability. It had the opposite effect in other patients, who would become subdued, apathetic, passive, and lacking in spontaneity. The harshest critics said that a lobotomy turns "lunatics into idiots."

Nonetheless, lobotomy is part of the history of psychiatry. In fact, it is part of the history of medicine itself – as a bad example of a momentary feeling of omnipotence in physicians.

Story 7.3: A French thinker and his lytic cocktail

The use of substances that affect the human mind is as old as humanity itself. Natural compounds that calm the mind and subdue passions have been known since ancient times. Some of those include the alkaloids atropine, scopolamine, and hyoscyamine that are found in a plant with the scary-sounding name of *deadly nightshade*. The Latin name is the more benign-sounding *Atropa belladonna.* The second part of the name, belladonna (also bella donna) is Italian for "beautiful lady." The name comes from the use of the plant to prepare eye drops for dilating the pupils, which would make women appear more seductive while also giving them blurry vision. Atropine is still used in medicine to dilate the pupils.

There is also opium, which was made from poppy seeds, and morphine was later isolated from opium. The sedative and hypnotic effects of cannabis (*Cannabis sativa*) were also known, as was one of the most used and still frequently recommended substances to alleviate mental tension: alcohol. In the 19th and early 20th century, they started to use the effects of those substances with specific intent; if not directly as treatment then at least to alleviate the symptoms caused by mental disorders. In the previous story, we talked about the use of other chemical compounds – potassium bromide, chloral hydrate, insulin, camphor, and barbiturates – but there were many more substances that were used to treat patients with mental disorders. Not one of them, however, has had such far-reaching implications for the history of psychiatry as chlorpromazine.

This book has already presented various research scientists whose activities and mindset far exceeded the boundaries of the initial area of their work. One such person is Henri Laborit, physician, writer, and philosopher. He began as a surgeon, but he was best known for physiology, pharmacology, ethology, and social sciences. He is one of the founders of modern anesthesiology and neuropsychopharmacology and coined the term *eutonology*, which refers to "adequate tone in all biological functions." Because of his multifaceted talent, historians consider him to be one of the leading French thinkers of the 20th century. Laborit was one of the few scientists to be honored and recognized on both sides of the Iron Curtain, in both the United States and the Soviet Union. Apart from all that, he was also the first person to recognize the potential of substance RP4560: chlorpromazine. As with so many great discoveries, he also came across this completely by accident.

Henri Laborit was born in 1914 in Hanoi, which at the time was the capital of French Indochina. He moved to France as a young boy, and was educated at the Naval Health Service. He then worked as a navy physician in another French colony, Tunisia.

As a military surgeon, he encountered the shock suffered by the body following severe injury and bleeding right from the beginning of his practice. Patients did not necessarily die as the result of their injuries, but as a result of the shock. If shock could be prevented or reduced, it would significantly contribute to the success of surgery. Lowering the body temperature of patients, a kind of hibernation, had proven to be a fairly successful method of reducing the likelihood of shock, but frostbite was a frequent adverse effect of this and the method was gradually abandoned.

Laborit began his experiments to prevent shock just after the end of World War II, while he was still in Tunisia. He used combinations of various anesthetics and hypnotics to pre-medicate patients before surgery and his combination later became known as the Laborit or lytic cocktail. The word lytic is derived from the Greek *lysis*, which can be translated as disintegration. In this case, the idea was to "disintegrate" the shock response. The cocktail really did work and reduced patients' shock response, but at the time it was considered too heretic, and other doctors did not trust him.

Laborit's experiments began to make headway when he was transferred from a minor military hospital to the physiology laboratory of the large Val-de-Grace military hospital in Paris. Here, he could fully concentrate on solving the shock response. As he had not been completely happy with the substances he had been using in his cocktail, he visited the pharmaceutical company Rhône-Poulenc and asked them for a sample of the new antihistamine labeled RP4560 that had been developed recently in the corporate laboratory by a research group headed by the chemist Paul Charpentier. It was Charpentier who later came up with the name that made this substance famous: chlorpromazine.

When Henri Laborit first used chlorpromazine in surgery, he was pleased to learn that the new compound contributed to reducing shock. To his surprise, though, the patients became relaxed and indifferent and were better able to handle the stressful pre-op procedures. He correctly suspected that these properties meant chlorpromazine could be effective in psychiatry. In order to evaluate those properties further, he asked his younger colleague, psychiatrist Cornelia Quarti, to test it on herself. Cornelia agreed, and the chlorpromazine worked. She slowly became weakened and lethargic, and the experiment ended with her passing out. That was the end of the testing of the new substance, however, because the hospital director banned any further experiments.

Chlorpromazine was applied for the first time on January 19, 1952. Jacques L was the first patient, and his acute mania was the first diagnosis. Mania, according to the American Psychiatric Association's diagnostic manual, is a "distinct period of abnormally and persistently elevated, expansive, or irritable mood and abnormally and persistently increased activity or energy".

The outlook for this patient was not too promising at first. However, on February 7, after three weeks of therapy, the patient was so calm that he played cards with the other patients in his room.

But that was not the watershed moment. Jacques L received chlorpromazine as part of the Laborit cocktail, along with two other substances. In addition to the drugs, the patient was also given hypothermia therapy to increase the effect. Electroconvulsive therapy was also used to ensure the results. It was unclear how great a role chlorpromazine had played in the patient's recovery, but even this result was enough for the medical circles in Paris to start whispering about a great new psychiatric drug.

Chlorpromazine was not the last significant contribution of Henri Laborit's medical career. He participated in research on the toxicity of oxygen, co-developed another two psychopharmaceutical drugs, and his contribution to the theory of human behavior was also significant. He developed the theory that people's mental problems are caused by their inability to adapt their instinctive responses to the modern social environment.

Laborit was nominated for the Nobel Prize for his work, but he did not win it. It is said that the reason was because the Parisian scientific community rejected his innovative principles. One of his opponents actually went to Stockholm to stop the jury from naming Henri Laborit as the laureate.

Story 7.4: A professor, his assistant, and psychiatric penicillin

There was a strict hierarchy at work at the Clinic of Mental Illnesses and the Brain (*Clinique des Maladies Mentales et de l'Encéphale*, or CMME) at the Saint-Anne Hospital in Paris. At the top of the hierarchical pyramid was the department chair and below him were two to three senior assistants. Below them were senior interns, even lower was a group of junior interns and at the bottom were the nurses and attendants. The hierarchy was so strictly observed that those at the higher levels often knew nothing of the work of their colleagues at the lower levels. Practically all decisions in the department were made by the man at the top of the pyramid.

In the mid-20th century, that man was Jean Delay. He was born in 1907 in the Basque town of Bayonne in southwestern France and he was a precocious child. He enrolled in medical school at the age of 18 and alongside medicine, he studied esthetics at the Sorbonne. He said that he chose psychiatry as a specialty because he hated the sight of blood. In his spare time, he wrote short stories and earned a doctorate in literature and philosophy. His dissertation supervisor was so happy with his philosophy dissertation that he tried to persuade Delay to leave medicine and focus instead on philosophy.

By the time he was 38, Delay was so well-known as a psychiatrist that he was appointed forensic psychiatrist to examine the war criminals at the Nuremberg trials. At the age of 52, he became a member of the French Academy (*Académie française*), a group of 40 preeminent intellectuals in France. He took the place of the deceased Louis Pasteur and after his own death, he was replaced by the oceanographer Jacques Cousteau.

Delay became a professor of medicine at 35, the youngest age ever for a professor of medicine in France at the time. Four years later, when World War II ended, he became the chair of CMME and took his place at the top of the hierarchical pyramid.

The second senior assistant at the clinic, essentially the third man in the hierarchy, was Pierre Deniker, a native of Paris and ten years younger than his boss. He joined CMME in 1939, before Jean Delay started working there. His work at the clinic was interrupted by World War II, during which he volunteered for the Red Cross and later became a member of the Free French Air Forces. He was awarded the Military Cross.

The first information about chlorpromazine came to the CMME from Deniker's brother-in-law, who was an anesthesiologist and had probably tried the Laborit cocktail containing chlorpromazine to calm patients before surgery. Pierre Deniker decided to try chlorpromazine by itself, under the supervision of his boss. He removed the other substances in the cocktail described by Laborit, but kept the hypothermia procedure. On one occasion, the pharmacy did not provide the necessary amount of ice, so one of the nurses decided to administer the drug without hypothermia. There was no reduction at all in the effectiveness of the chlorpromazine.

As with the earlier cases, we know the name of the first patient to receive chlorpromazine therapy: Giovanni A, a 57-year-old laborer with a long history of mental illness. He was admitted to the clinic for "*making improvised political speeches in cafés, being involved in fights with strangers, and walking around the streets with a pot of flowers on his head proclaiming his love of liberty*." Giovanni A received his first dose of chlorpromazine on March 24, 1952. Nine days later he began conversing normally and after three weeks he was discharged from the hospital. The treatment of another seven patients, and then dozens of others, was equally successful. The results were more than just groundbreaking. Until that time, manic patients were the least popular in psychiatric clinics because it was so difficult to help them. To keep them from hurting themselves and other patients, they often had to be restrained with straitjackets or had their hands shackled to their beds. As of March 1952, psychiatrists had a means to help these patients for the first time. Chlorpromazine was soon successfully used to treat not only acute mental illnesses but also as a therapy for chronic, long-term disorders. Schizophrenics who had spent years locked up in mental institutions showed no symptoms of their disease after treatment with chlorpromazine. The drug was not only more effective than the previous therapies presented in our earlier stories, it was also safer, better tolerated, and a much simpler alternative.

A new era in the history of psychiatry began with the discovery of chlorpromazine: the era of psychopharmacology. The introduction of chlorpromazine as a treatment for mental disorders is considered just as much of a gamechanger as the first use of penicillin in the treatment of infectious diseases. In fact, some historians call chlorpromazine the psychiatric penicillin.

When considering the significance of both, there is nothing wrong with that comparison. There is, however, one enormous difference between penicillin and chlorpromazine. The effects of penicillin (and its predecessors) were only discovered after the causes of the diseases they were looking to treat were determined. The microorganisms that cause infectious diseases were identified, isolated, and described in detail before their eradicators were discovered. That means penicillin and its antibiotic predecessors were and are purposefully developed, even if with a certain dose of serendipity, to fight a known foe: harmful microorganisms.

Conversely, the causes of schizophrenia and other major and minor mental illnesses were completely unknown at the time that chlorpromazine was discovered. The discovery of chlorpromazine and its potential use in psychiatry was completely accidental. Despite all the efforts of psychiatrists and experts in other fields, the cause of most mental illness remains unknown more than half a century after chlorpromazine was first used, even with the enormous amount of information that has since been gathered. The differences in the brains of schizophrenics and healthy people have been described and the factors that increase the likelihood of mental illness are known, but the causes of those illnesses remain unknown. While penicillin is a causal therapy – that is, it works to eliminate the *cause* of the illness – chlorpromazine is a symptomatic treatment that works to eliminate the *symptoms* of the illness.

Chlorpromazine was the first drug of a class that was later given the name antipsychotics, because it works on a group of mental disorders called psychoses.

Psychosis is a serious mental illness in which patients lose touch with reality, perceiving and judging things differently from others while not recognizing the change in themselves. Schizophrenia and bipolar disorder are both forms of psychosis.

Regardless of how chlorpromazine worked, the fact is it did work and it helped patients, and that makes its discovery a real revolution in psychiatry. As one historian enthusiastically wrote, "*...the atmosphere in the disturbed wards of mental hospitals...was transformed: straitjackets, psychohydraulic packs and noise were things of the past! Once more, Paris psychiatrists, who long ago unchained the chained, became pioneers in liberating their patients, this time from inner torments... It accomplished the pharmacologic revolution of psychiatry.*"

Story 7.5: Psychoanalysis and the need to know foreign languages

News of the success of the Parisian psychiatrists spread quickly, and after the first reports by Jean Delay and Pierre Deniker were published, chlorpromazine treatment began to make its way around the world. In 1955, three years after the success in the first patient, results had been collected from psychiatrists in Switzerland, England, Germany, Hungary, the Soviet Union, Canada, Latin America, and the U.S.

It was not all that easy in the U.S., however. In the 1950s, psychiatry in America was almost completely limited to psychoanalysis. To understand why that was and what it meant for the development of psychiatry in general, and psychopharmacology in particular, we need to go back half a century. We also need to move a few thousand miles to the east, to the cradle of psychoanalysis: central Europe in the late 19th and early 20th century.

Viennese neurologist Sigmund Freud, born Sigismund Schlomo Freud in the Moravian town of Freiberg, is considered the father of psychoanalysis, and rightfully so. His theory of psychoanalysis was based on the principle that mental disorders in adulthood, particularly neuroses, were caused by repressed sexual memories and dreams from childhood. This initial concept was later expanded and the basic principles can be summarized like this: Human attitudes, behavior, emotions, and motives are to a great degree influenced by irrational processes that are rooted in the subconscious mind and originate in forgotten or repressed childhood events. Mental and emotional disorders are then the result of a conflict between the conscious and the repressed memories in the subconscious.

Psychoanalysis was initially more of a theory than a therapeutic method. Only through purposeful guidance were elements of the subconscious liberated and brought into the conscious mind, making it possible to eliminate the consequences of those elements: mental disorders. This concept and therapeutic approach worked fairly well for neuroses, neurotic tendencies, fears, and minor depression.

Psychoanalysis contributed greatly to the advancement of psychiatry, by allowing psychiatrists to go beyond the walls of hospitals and asylums. They could see patients in their own offices, as doctors with standard specializations could. The problem was that the main patients, the ones with schizophrenia, manic disorders and severe depression, still had to be contained within the walls of institutions. It is easy to understand why one of the early opponents of psychoanalysis said that, as a therapeutic method, it sooner helped the misfortunate than the insane.

Before World War II, psychoanalysis was more or less strictly the domain of physicians in Europe, mainly in central Europe. After the war, the situation began to change swiftly and soon the U.S. became the epicenter of psychoanalysis. Before the war, American psychiatry was still biological psychiatry; after the war, it slowly became psychoanalytical psychiatry. By the 1960s, psychoanalysis had practically squeezed out biological psychiatry and it had become synonymous with psychiatry itself. The main goal was still primarily to treat neuroses and depression but gradually, as it grew in popularity, some American psychiatrists tried to use psychoanalysis to explain and treat severe mental illness, including psychoses. The following explanations offered by some American psychoanalysts are an example of how mental illness was viewed from the psychoanalytical perspective: schizophrenia was a failed response to fear experienced in childhood, particularly fear caused by one's mother; depression was explained as an unanswered plea for love; paranoia was formed in the first six months during breastfeeding and was caused by a warped relationship between mother and child.

If this was how mental illness was understood, it is not hard to see how American psychiatrists could reject chemical treatment. It is almost impossible to reconcile the belief that abnormal behavior is caused by the faulty behavior of a mother toward her infant with the belief that the same illness is caused by the faulty behavior of chemicals in the human brain. That would all soon change, and the first step towards that change was made by a Jewish doctor who fled Germany, settled in Montreal, and learned to speak French from his French-Canadian wife. All of this – the doctor's background, Montreal, and the French language – played an important role in the history of chlorpromazine.

That Heinz Lehmann (born Heinz Edgar Lehmann) was an extraordinary man can be seen not only in the kind of physician he was, but in the kind of person he was in private and in public life. He was from a family with a medical tradition going back generations, which made his future more than certain. Lehmann read all of Freud's work when he was 14 and he was so affected by it that he decided to become a psychiatrist. He studied medicine at several German universities and at the center for psychoanalysis in Vienna. Like thousands of other people of his heritage, with the rise of the Nazi regime he decided to emigrate, choosing Canada as his new home. In order to get permission to leave, he said he was going on a skiing trip to Quebec. In 1937, he found himself in Canada with a pair of skis propped on his shoulder and a bag packed for a two-week visit.

A few months later, he had found work in a psychiatric hospital in Montreal, where he was responsible for a ward of 600 patients. He remained there for a full

35 years, and as he himself said, it was his best university. He married and, with the help of his wife, became trilingual, learning to speak fluent French in addition to his native German and then English.

Lehmann was known for his hard work, humility, and human approach. He would typically work late into the night, with his workday ending just before dawn when the hospital's pharmacy was preparing medications for the upcoming day. He lived with his family in a house on the hospital grounds, far removed from the city center. He never owned a car and he went everywhere by bicycle. When he became the hospital director in 1947, he implemented a custom that he continued until his death: every Christmas, he would go around to all the patients and employees, personally wishing each of them a happy holiday. One of his students measured Lehmann's "Christmas trip" around the hospital – it was eight miles.

In the late 1960s, Heinz Lehmann made efforts to alleviate the tense atmosphere during the anti-psychiatry movement and it was he who advocated for the International College of Neuropsychopharmacology congress to be held in Prague in 1968 – not despite the occupation of Czechoslovakia by Soviet tanks, but because of it.

He encountered chlorpromazine in the spring of 1953, shortly after Delay and Deniker published their first report. Lehmann was fortunate in that Montreal, where he worked, was the Canadian home of the French pharmaceutical company Rhône-Poulenc. This was the same company that had developed chlorpromazine, which it marketed under the name Largactil. The company's representatives did not have to go far to get to the psychiatric hospital. Besides, it seemed that Heinz Lehmann was the only one in the area to whom literature about chlorpromazine would make any sense, since he was the only one who knew how to read scientific texts in French. And so, one spring Sunday as he was soaking in the tub, he read a few articles written by some French colleagues. They had been dropped off a few days earlier by a diligent representative from the pharmaceutical company. Lehmann was so impressed by what he read that the next Monday he began to organize a clinical trial with 72 patients. The results were so surprising in the first few weeks that at first he thought it was just a fluke. That the patients would stop hallucinating and being delusional after just a few days was extraordinary at the time. In August, after the clinical trial was completed, a large percentage of the patients showed no signs of mental illness.

Lehmann's language skills came in handy again four years later. In the airplane on his way home from the International Congress of Psychiatry in Europe in 1957, he was reading some of the handouts from the congress. An article in German about a new antidepressant caught his attention and after he returned to Montreal, he started to use the new drug – imipramine – to treat his patients. He was the first psychiatrist in North America to use the drug.

His results with chlorpromazine were published in 1954 in a prestigious psychiatric journal. That same year, those results were presented for the first time on

American soil at the annual conference of the American Psychiatric Association. Obviously, one speech could not cause a great shift in the psychiatric paradigm. Psychoanalysis as a concept and method of therapy was the preferred approach in the U.S., despite Heinz Lehmann's unequivocal findings. But he did convey the results from the French psychiatrists to the American continent and was the first to leave a psychopharmacological imprint.

Despite his unquestionable contribution to psychopharmacology, Heinz Lehmann was never a blind follower of it. He commented in 1995 that "*the cookbook approach offered by drug therapy is easy and doesn't require a lot of thought.*" He said that many seemed to think that listening to patients was outdated, but he believed that what at that time looked antiquated would one day make a comeback.

Concluding remarks

While the majority of contemporary psychiatrists believe that chlorpromazine was a real gamechanger in the treatment of severe mental illness and the beginning of the psychopharmacological revolution, the history of the drug is not all positive. The drug itself was never able to solve all the problems associated with the treatment of schizophrenia and other psychoses, and to this day doctors still do not fully understand its mechanism of action. Furthermore, chlorpromazine is a drug, and as such, it has adverse side effects to go along with its benefits. Some of the most serious of these include a shuffling gait, uncontrollable facial muscle and other body movements, as well as confusion.

These were the side effects that made it a symbol of the anti-psychiatry movement of the late 1960s. Opponents claimed that giving drugs to patients victimized them and actually did nothing to help them. They said the drugs were just replacing the old physical straitjackets with chemical ones. Chlorpromazine was dubbed the "chemical lobotomy." The anti-psychiatry movement even went so far as to demand psychiatry be eliminated as a medical specialization. The fate of Professor Jean Delay, one of the discoverers of chlorpromazine, shows us that this was not just some academic theorizing. During the revolutionary events of 1968 in France, hundreds of people, mostly students, broke into his offices and ransacked them. Jean Delay was not the only one to experience for himself the hatred of the anti-psychiatry movement. Even the affable Heinz Lehmann was targeted. During a public debate, someone threw a piece of cake in his face. Another leading psychiatrist, Herman van Praag, actually had to seek police protection for himself and his family.

Although the intensity of these opinions has lessened over time, anti-psychiatry is still alive and well and complicating the treatment of those patients who really need it.

Whatever the negative attributes of chlorpromazine may be, the positives greatly outweigh them. It was the first drug to demonstrate that the symptoms of even severe mental illness can be managed. The significance of chlorpromazine does not end with therapy. It was a drug that stimulated research of the biological causes of mental illness, and it started a psychopharmacological revolution that changed how psychiatry as a medical specialization is seen, turning it into a full-fledged medical field.

But most importantly, chlorpromazine caused a major change in the approach used with psychiatric patients, which can be summarized in one word: deinstitutionalization. Illnesses that used to require locking patients up in mental institutions could now be treated in a physician's office without the need for isolating the patient. Patients could lead a normal, and very often a full, life in normal society. Something like that was unheard of before the discovery of chlorpromazine. The most significant changes took place in the U.S. In 1955, there were 560,000 patients housed in mental institutions. That number slowly declined to 337,000 in 1960; to 150,000 in 1980; to 120,000 in 1990; and to less than 30,000 in 2004.

Today, chlorpromazine is psychopharmacological history, having now been replaced by more effective, safer, and better-tolerated drugs. But nobody can ever take away its place as the cornerstone of modern psychiatry.

8

Prozac

"Depression is a lot like that: Slowly, over the years, the data will accumulate in your heart and mind, a computer program for total negativity will build into your system, make life feel more and more unbearable. But you won't even notice it coming on, thinking it is somehow normal, something about getting older, about turning eight, or turning twelve, or turning fifteen, and then one day you realize that your entire life is just awful, not worth living, a horror and a black blot on the white terrain of human existence. One morning you wake up afraid you are going to live."

That is an excerpt from the book *Prozac Nation* by American Elizabeth Wurtzel, in which she describes her experience with depression. It is a frank and very personal account of how the writer falls deeper and deeper into the spiral of depression and how the illness jeopardizes not only her college ambitions, but her entire life as well. It also describes how it affected her relationships – with men, her family, and friends – and culminated in a suicide attempt. But the book is mainly about how the young woman was finally able to free herself from the grips of depression with the help of the antidepressant Prozac.[1]

Although reviews were highly critical, the book was popular with readers and became an (almost) instant bestseller. Ten editions were printed and it was translated into 15 languages. It influenced an entire generation of young people, mainly in the U.S. Along with the drug which adorns both the book cover and the title of this chapter, it helped to change how society views the illness called depression.

[1] We have the same terminological dilemma here as we did with Aspirin/aspirin. All drugs should be given under their generic names, which in this case is fluoxetine. However, since the entire world knows it only as Prozac, that is the name used in this chapter.

V. Marko, *From Aspirin to Viagra*, Springer Praxis Books,
https://doi.org/10.1007/978-3-030-44286-6_8

Story 8.1: Two psychiatrists, a singing cyclist, and dancing patients

In the mid-20th century, the time when the story of our next drug begins, opinions about psychiatry were slowly starting to change. It stopped being something like a social field and started transforming into a full-fledged biological and medical field. Biological psychiatry started to see the development of a new subfield, that of psychopharmacology. Barbiturates had long been known (we mentioned them in the chapter about chlorpromazine), and the first information about lithium treatment was beginning to appear. But the watershed moment came with the anxiolytic drug, meprobamate. It was the first drug that psychiatrists could prescribe in their daily practice, and prescribe it they did. Meprobamate was introduced to the U.S. market in 1955 and within a year, one in every 20 Americans was using it monthly. It was the first psychopharmaceutical to break the boundaries of pharmacology, long before Prozac, and became a part of the history of American culture.

That was just the beginning. A few years later, a new class of drugs was introduced – benzodiazepines – that would long dominate both the pharmaceutical market and the minds of patients, especially those living in the U.S. It was a drug from this class, known as diazepam, that thrilled both psychiatrists and their patients. Diazepam became a part of every family's first-aid kit and by 1970, every fifth woman in America was using it. Benzodiazepines accelerated the revolution in psychiatry, and a field that had been oriented toward the mind (and psyche) of patients was slowly becoming a field that studied the changes in the chemical processes in the brain.

That was still only the beginning, however. The shift that began with the discovery of chlorpromazine, and was furthered when benzodiazepines were introduced, came to an end with the discovery and introduction of antidepressants. Two psychiatrists stood at the beginning of that end: Roland Kuhn and Nathan Kline. They were separated by geography and character and by their approach to psychopharmacology.

Roland Kuhn was born in 1912 in Biel/Bienne in the canton of Bern in Switzerland. He started out wanting to be a surgeon, but this specialization was not offered at Bern University and he ended up studying psychiatry under Jakob Klaesi – the one who had introduced sleep treatment for mental disorders and whom we met in one of the stories in the chapter on chlorpromazine. After graduation, Roland Kuhn took a job at the Münsterlingen psychiatric hospital, on the premises of an old manor house on the shores of Lake Constance, where he worked until he retired and where he made the majority of his clinical discoveries.

But Roland Kuhn was first and foremost a philosopher. Most of his publications concerned various areas of psychiatry and psychology and how they are linked with philosophy, sociology, ethnology, esthetics, and art. A colloquium was held in celebration of his 80th birthday, where the main themes were ethics and esthetics in philosophy and psychiatry.

While psychopharmacology was not Roland Kuhn's main area of work, it was the discoveries he made in this emerging field of science that made him famous. Like many psychiatrists of the time, he had heard about the French psychiatrists Delay and Deniker and their work with the antipsychotic drug chlorpromazine, and he used that drug to treat his patients. However, the hospital had a tight budget that was insufficient to buy large amounts of the new drug. In order to continue with his trials, Kuhn asked the pharmaceutical company Geigy to provide a substance with similar effects, and one of the substances he got in return had the code name G22355. It was similar in structure to chlorpromazine and it was presumed to have similar sedative effects. The results were the exact opposite of the expectations. The drug did not calm the patients down and instead, it made them even more agitated and manic. When one of the patients escaped from the hospital on a bicycle and rode around town in his nightshirt singing at the top of his lungs, the trial was immediately stopped.

Kuhn and his coworkers evaluated the effects of the new drug and decided they would not try it on patients who were psychotic or manic. Instead, they would try it on patients with depression. Its effects upon the first three patients were so dramatic that they knew right away they had found a drug for depression. This is another case where we know the name of the first patient: one Paula J.F, who suffered from depression. Her treatment began on January 12, 1956. Six days later, on January 18, she was "*totally changed*."

Despite the importance of Kuhn's discovery, neither Geigy nor the professional community accepted the antidepressant effects of imipramine, as Geigy called the new drug. Roland Kuhn published his report in a Swiss journal in August 1957. A month later he presented it at the 2nd international congress for psychiatry, and in November 1958 it appeared in a top American medical journal. There was practically no response. Participants said that not even a dozen people came to his speech at the psychiatry congress. However, Heinz Lehmann read up on the speeches on his way from the congress back home to Canada. You will remember Heinz Lehmann from the chapter on chlorpromazine. He was the very first to try the new antidepressant on the American continent.

Imipramine was introduced to the Swiss market in 1957, and other European markets followed a year later. Still there was not much feedback. Its practical use started to spread in the 1960s after the drug was given a huge push by two events. The first was that the wife of Robert Böhringer – one of the largest shareholders of Geigy – became depressed. Böhringer asked for a sample of the new antidepressant, took it home and his wife recovered. It goes without saying that the weight of his shareholding in the company pushed development into high gear.

The second event that sped up the spread of imipramine was the discovery of another antidepressant, called iproniazid, on the other side of the Atlantic, which was getting rave reviews from physicians and the public alike.

That brings us to the discoverer of the other antidepressant, Nathan S. Kline. Unlike his Swiss colleague, who was humble, pedantic, and extremely consistent in his opinions, Nathan Schellenberg Kline was a dynamic and charismatic man who was always open to new ideas. And while Roland Kuhn was, and remained, a country doctor, Nathan S. Kline was a notable director at the Rockland Research Institute and holder of two prestigious Lasker awards, which are given annually in the U.S. for major contributions to medical research. Upon his death, the *Rockland Research Institute* was renamed the *Nathan Kline Institute for Psychiatric Research.*

Before we get to the antidepressant that Nathan Kline is credited with discovering, let us go back a few years to the World War II era, and to the German V-2 rocket. The fuel used in the first rocket systems was a mixture of ethanol and liquid oxygen. Towards the end of the war when the sources of this fuel were diminishing, German engineers began using a substance known as hydrazine. Large supplies of hydrazine remained after the war ended and ways to put the substance to use were being researched. One of the potential uses was in the chemical and pharmaceutical industries. The chemists at Hoffmann-La Roche (Roche, for short) in New Jersey seem to have taken this the furthest. They prepared a substance from hydrazine that they called iproniazid, which they tested in 1952, along with a similar substance, as a treatment for tuberculosis. Following World War II, tuberculosis was a widespread disease and many new drugs to treat it were being tested.

The good news was that both substances proved to be effective, and additionally, they improved the patients' mood (particularly iproniazid). On March 3, 1952, *Life Magazine* featured an article with photos of smiling tuberculosis patients dancing in the hallways of hospital wards.

Later events clearly document how differently the treatment of depression and other mood disorders was viewed in the mid-1950s and the huge advancements that were made in psychopharmaceuticals in the following years. In the case of iproniazid, the fact that it made patients happy was seen more as an adverse side effect. No further research was done on this property of the drug and it was classified as an anti-tuberculosis drug. There were attempts in 1956 to promote iproniazid as an antidepressant, but to no avail. In fact, Roche did not see any future in this field.

Around this time, 40-year-old Nathan S. Kline appeared on the scene. He asked Roche about the possibility of trying iproniazid on psychiatric patients, and although the reply was unfavorable – Roche had no interest in antidepressants – Kline and his colleagues tested it successfully on several patients anyway. Two-thirds of the patients responded to treatment with a better mood and increased activity, but it was still not enough to change the company's mind, even after a personal presentation of the results to the president of the company. Kline and his group of people decided to speak directly to the public. There was a story in the

New York Times in April 1957 entitled *Mental Drug Shows Promise*. More psychiatrists were learning about the new drug and an increasing number of patients were enjoying the benefits. Although it was withdrawn from the market several years later for its adverse effects, medical historians still consider it a major contribution to the treatment of depression.

In the U.S., iproniazid was marketed under the name Marsilid. At an American racetrack on September 4, 1959, a horse named Marsilid won the race; it seems likely that the owner of the horse was a patient who had been successfully treated with the drug.

Despite the fact that the discovery of iproniazid was the product of teamwork, Nathan S. Kline alone received the Lasker Prize. What followed was a bitter fight, not the first or last in science, for recognition of the contribution made by two other colleagues. The fight not only played out in the bullpen at scientific journals and at conferences, it was also taken to court. After several years of litigation, the others who were seeking recognition finally got what they wanted. Nonetheless, Nathan S. Kline is still known as the official discoverer of iproniazid, one of the first two antidepressants.

Story 8.2: Three chemists and three neurotransmitters

To understand the history of Prozac and its immense success, we have to look at the prevailing theoretical principles of psychopharmacology in the late 20th century.

We will start with how nerve cells (or neurons) communicate with one another and how, because of that communication, signals pass from one neuron to the next and on throughout the entire nerve. All the processes in the brain, including thoughts and feelings, are caused by the ability of neurons to pass along information across small and larger distances.

Each neuron has several long tendrils known as axons, and these are responsible for sending the signal. The connection between the axons that communicate with nerve cells is not direct and fixed – there is a small gap called a synapse or synaptic cleft. Chemicals called neurotransmitters are responsible for transmitting the signal from the nerve ending of one neuron over the synapse to the nerve ending of another neuron. They are usually formed directly in the nerve endings where they cluster in small spherical cells. When a signal is received by a nerve ending, the neurotransmitters are released into the synapse. They pass through to the nerve ending on the adjacent neuron, where they incite the continuation of the signal. The signal passes through the second neuron until it reaches the other end, where the process is repeated, and in this way the signal passes through the entire nerve from one neuron to another. Several neurotransmitters take part in transmitting the signal, but for us there are four that are important: serotonin, noradrenaline, dopamine, and acetylcholine.

As long as the required amount of the relevant neurotransmitter is available in the synapses, everything runs along smoothly and the nerve signal is spread through the brain at the necessary intensity. However, as soon as the balance of neurotransmitters is thrown out of whack for whatever reason, usually due to a deficiency in one of them, it becomes a problem. This neurotransmitter imbalance results in mental disorders.

Over time, research papers began to appear that linked a specific mental disorder to a deficiency of a specific neurotransmitter. In the 1970s, there was an agreement on the assumption that one neurotransmitter equals one illness. A deficiency of noradrenaline and serotonin was responsible for depression and other mood disorders; too much dopamine caused schizophrenia; and a too little acetylcholine was paired with dementia.

The mono-transmitter theory or model is today considered to be too simplistic and out of line with the latest knowledge. However, at the time it was disseminated, the psychopharmacological and psychiatric ramifications were far-reaching. For the first time, a mental illness could be linked to a specific biological problem. For the first time, the relationship of cause and effect could be applied to psychiatry – unless, of course, you take into account the psychoanalytical relationship between repressed negative memories from childhood and severe mental illness in adulthood. Also for the first time, drugs could be synthesized – aimed at a specific biological target – that could be effective against a specific mental disorder. If a drug could make more serotonin available in the synapses, an antidepressant effect could be expected; if it could reduce the amount of dopamine, the effect would be antipsychotic.

The drug that very selectively raised the concentration of serotonin in the synapses was fluoxetine, better known by the name it is marketed under: Prozac.

When the first antipsychotics and antidepressants were discovered and doctors began prescribing them for patients, practically none of this science was known. Nothing was known about how they worked, and in fact very little was known about the chemistry of the brain in general. The prevailing theory in the 1930s and 1940s was that the transport of signals and information in the brain was executed by electric transmission. The nerve cells were considered akin to power lines that transmitted electrical signals from one place to another without any interruption. The idea that variances in the concentration of certain chemicals in the brain may be responsible for behavioral changes began to appear in the 1950s. It was a radical concept at the time, but it took hold and dominated psychiatrics for the next 50 years and it still prevails today.

Many people deserve to be mentioned here, but the three that receive the most mention are Bernard B. Brodie, Julius Axelrod, and Arvid Carlsson. Each was from a different part of the world and each had a different personality. Fate would have it that they found themselves in the same place, however briefly, at the chemical pharmacology laboratory of the National Heart Institute in Bethesda, Maryland, part of the National Institutes of Health. It was in this laboratory that the cornerstone was laid for the neurotransmitter theory of mental illness. Essentially, each of those men is responsible for the description of the properties of one neurotransmitter: Brodie for serotonin, Axelrod for noradrenaline, and Carlsson for dopamine.

The first of the three, Bernard Beryl (Steve) Brodie, was a far cry from the usual image of scientists as boring, self-absorbed introverts. He was nicknamed Steve in honor of adventurer Steve Brodie, a New Yorker who jumped off the Brooklyn Bridge in 1886 on a $200 bet. The nickname was supposed to convey his ability to take risks and take advantage of every opportunity that came his way.

Brodie was born in Liverpool, England, on August 7, 1907. He said of himself, with a smile, that he developed slowly as a child. He did not learn to talk until he was four, and he did not stand out in high school much either. One of his teachers refused to recommend him for a summer job, while he also had a disagreement with the principle and dropped out of school, opting to enlist in the army instead. Fortunately for him. The army transformed a shy teenager into a confident and courageous young man who later won the Canadian Army boxing championship, although he said that it was not because he wanted to win in boxing but because he wanted to keep from getting hurt by his adversary.

In the meantime, his family moved to Canada, where his father had a small fabric shop in Ottawa. When Brodie was discharged from the military, he enrolled at McGill University in Montreal, earning money for tuition by playing poker. At first, it looked like even college was not going to go all that well for him and his career as a scientist. He even fell asleep once during a chemistry class. Things finally changed for Bernard B. Brodie in his fourth year, when he assisted a chemistry professor with an experiment. He ultimately graduated and began his science career.

The reason Brodie is part of this story is that he significantly contributed to the creation of the neurotransmitter theory of nerve signal transmission. He was already well into his science career when he forayed into this subject, and he had become a well-known pharmacologist much earlier. Brodie contributed to the discovery of two drugs that have nothing at all to do with neurotransmission. Coincidentally, both drugs have something in common with the drugs mentioned in earlier chapters. The first was the successor to quinine, Atabrine, and the second is the well-known and still marketed acetaminophen, or paracetamol. This drug was mentioned in the chapter about aspirin. More important than Bernard B. Brodie's discovery of new drugs, however, was his contribution to the establishment of a new field of science called pharmacokinetics, a branch of pharmacology that deals with the fate of drugs within the body. That is, how the drug is distributed and metabolized in the body and how it is excreted from the body.

Next, we have Julius Axelrod. The son of Jewish immigrants from Poland, he was born on May 30, 1912 in Manhattan's Lower East Side, in an impoverished neighborhood of mostly Jewish immigrants. His father earned money weaving baskets. Just like his older colleague, Julius was not an exceptional student and he therefore did not make the quotas for Jewish student enrollment in medical school. He had to go to night school, finally graduating with "only" a degree in chemistry at the age of 29. He lost also his left eye in the explosion of an ammonia bottle.

His lucky day arrived in 1946, when his collaboration with Bernard Brodie began. Brodie became Axelrod's mentor and they worked together on many projects, including the drugs Atabrine and acetaminophen mentioned earlier. Together they went to work for the newly-established National Heart Institute, where they began working on nerve signal transmission in the 1950s. While Brodie preferred to work with serotonin as the substance responsible for mood change, Axelrod favored a different neurotransmitter: noradrenaline.

The last of the group, Arvid Carlsson, was the youngest of the three. He was born in 1923 in Uppsala, Sweden. In contrast to his two older colleagues who had grown up in poverty, he was born into a family of academicians; his father was a history professor. Carlsson spent his college years at the prestigious Lund University. The years he spent at school overlapped with World War II, and although Sweden was neutral in the war, medical student Arvid Carlsson did encounter its consequences. In the spring of 1944, he was a member of a medical team which cared for refugees from the German concentration camps that had managed to get to Sweden.

Arvid Carlsson joined Bernard Brodie's team in August 1955 and immediately got to work on the substances that cause nerve signal transmission. Although he only spent about six months in Brodie's chemical pharmacology laboratory, it influenced his entire science career. Since two other neurotransmitters, serotonin and noradrenaline, had already been grabbed, Carlsson worked on the third, dopamine.

So now we have three scientists extraordinaire and their three neurotransmitters: Brodie's serotonin, Axelrod's noradrenaline, and Carlsson's dopamine. The Nobel Prizes received by both Julius Axelrod and Arvid Carlsson would indicate that their discoveries were truly revolutionary. Bernard B. (Steve) Brodie did not receive a Nobel Prize, despite being the first person to make the link between the deficiency of a specific chemical in the brain and a mental disorder. In more general terms, he was the first to discover that biochemical processes in the brain are responsible for changes in behavior.

But which neurotransmitter is the main and most important in transmitting nerve signals? A deficiency of which one causes mental and nervous disorders? As far as depression is concerned, at the time of the story and in the years following, serotonin was the winner. Because a serotonin deficiency in the synaptic cleft is the cause of depression, there was just one more little thing to do: find a chemical that was safe enough and tolerable that could stabilize serotonin in the synapses. That is where Prozac entered the picture.

Story 8.3: The role of the medicine box

The situation in psychiatry on one side of the Atlantic differed greatly from that on the other side prior to the invention of Prozac. In Europe, the diagnosis and treatment of mental illness was strictly linked to the cause of the illness and comparison of internal differences. To summarize, there were two types of depression. Endogenous (internal) depression had internal causes and symptoms that could not be directly linked to any external stimulus, whereas reactive depression could be traced back to an exact time and cause. Most often it involved sad events in the patient's life, such as the death of a loved one, a serious illness, or an unexpected divorce. The first type, endogenous depression, was very difficult to cure. Psychotherapy did not improve the patient's condition greatly, and neither did drugs. The most widespread treatment method – after its invention by "Maestro" Ugo Cerletti – was electroconvulsive therapy. We saw this in the chapter about chlorpromazine. If any treatment worked at all, it was limited to inpatient therapy in the wards of asylums and hospitals.

Conversely, the other type, reactive depression, could be managed with psychotherapy and with psychopharmacology, and often a combination of the two would be used. However, only patients with severe illness sought help, patients who were incapable of handling their situation on their own. In these cases, inpatient care was also often used. That meant that in Europe, psychiatric help was almost exclusively limited to mental hospitals and later, to the psychiatric wards of general hospitals. Outpatient psychiatry, which was often based on psychoanalysis, was rarely seen.

The situation in the United States in the 1960s and 1970s was different. When the principles and methods of psychoanalysis made their way from Europe to America just before and during World War II, many American psychiatrists embraced them. One reason for the quick acceptance was the possibility offered by psychoanalysis, in getting patients out of the asylums and onto the streets. Private psychiatric practices opened up right alongside the private practices of general, gynecological, and ophthalmological medicine. With the exception of a few cases, however, psychoanalysis mainly helped patients with minor mental illness. Even the founder of psychoanalysis himself, Sigmund Freud, used it mainly for neuroses. Moreover, the diagnosis of a mental illness was not necessary for psychoanalysis, because the deep-rooted cause of the illness was more important than the classification of symptoms. Even in this case, the majority of patients with severe mental illness were reliant on the help and long-term care found in mental asylums.

The number of patients treated for depression was limited to the truly severe cases. In the heyday of mental institutions, they were more of a horror show than a five-star hotel and everyone avoided them like the plague. The number of recognized and treated cases of depression was quite low; in fact, it was just the tip of the iceberg. Unsurprisingly, as we saw in the previous chapter, the pharmaceutical companies had little to no interest in treating this illness. Biological psychiatry in the U.S. began to make advancements after John F. Kennedy was elected president. His support for psychiatric research was most likely related to the mental health issues in his own family. His older sister Rosemary suffered from severe mental illness and was one of the first patients to undergo a lobotomy.

Before Prozac, the treatment of depression was more of an art than a science. The relationship between the cause of the illness and the treatment was unknown, but there were various strategies for specific types of patient and every doctor had their own "fool-proof" treatment method. Agitated and excitable patients were prescribed antidepressants with sedative effects (one tablet before bedtime). If a patient complained of low energy, they were prescribed antidepressants with stimulating effects (one tablet before breakfast). That would all soon change radically.

Many firsts are associated with Prozac. It was the first psychiatric drug, i.e. psychopharmaceutical, for which research and development was completely under the control of a pharmaceutical company. That company was *Eli Lilly and Company* (Eli Lilly). It was the first psychopharmaceutical invented through targeted synthesis, using models that simulated its future effects. It was developed as the result of a costly process through the combined efforts of scientists, using the cutting-edge technology of the time. That makes it difficult to point a finger at one inventor, or at least at the main inventor.

In the literal sense of the word, the "creator" of Prozac is chemist Bryan Molloy. Originally from Scotland, he was born in 1939 in Broughty Ferry (Bruach Tatha in Scottish Gaelic), a suburb of Dundee. After attending university in his native Scotland, he completed his education at Columbia University in New York, and in 1966 he took a job with Eli Lilly. The history of Prozac also begins somewhere around 1966, but the first compounds Molloy synthesized did not lead to Prozac. He developed them as potential drugs to treat heart disease. However, after a short while he took the advice of an older coworker and began synthesizing compounds that could affect neurotransmitters in the nerve synapses. He prepared 57 molecules of various structures, and these all found their way to David Wong. If we call Bryan Molloy the creator of Prozac, then we have to call David Wong its discoverer.

David T. Wong was born in Hong Kong in 1935. He was expected to follow in his father's footsteps after completing school and take over his successful machining business. It was not something he really wanted to do, especially after a machine he was working on once tore off his right thumbnail. David Wong stayed in school, first at National Taiwan University and then at universities in Oregon. It is often the little things that decide a person's fate. For young university graduate David Wong, it was the box of insulin he had seen as a little boy at his grandmother's. As with every package of medicine, it bore the logo of a pharmaceutical company. The words *Eli Lilly* were etched into his memory; so much so, that right after he finished school, he took a job with the company.

In the early 1970s, David Wong took the 57 different compounds from Bryan Molloy and began testing them on a new model system. This was a recently developed model that made it possible to evaluate the action of a compound on the nerve endings isolated from the brain of a rat. Compound number 82816 showed the best results in selective action on serotonin. After a minor tweak, it became compound number 110140 and that compound was later given the generic name fluoxetine. In 1976, it was also given the marketing name that it is known by the world over – Prozac.

Eli Lilly now had a compound which selectively raised the concentration of serotonin in the synaptic cleft, at least in the model system. However, the company did not know what it was good for and what practical use it had. The preliminary clinical studies showed that the new drug had very few side effects, so it made sense to pursue it further. At first, they wanted to use its potential antihypertensive properties, whereas in later clinical trials it was shown to have the side effect of reducing the patients' weight and they considered making this the primary property. They were looking at anything, just not depression. When a prominent British psychiatrist suggested that researchers at Eli Lilly use Prozac (still known then as fluoxetine) as an antidepressant, the response was crystal clear. There was no intention to use the drug for that purpose. However, further clinical studies provided more and more proof that depression was the area where Prozac would find its purpose. In 1987, the FDA registered Prozac as a new drug for the treatment of depression, 13 years after David T. Wong had described the beneficial properties of compound number 110140 in the model system and 30 years after the first antidepressants had been introduced. Every star has an exact date of birth. Prozac was born on December 29, 1987.

When *The New York Times* published a brief article in 1987 about a new drug for depression that would soon be approved, the price of Eli Lilly shares skyrocketed in just one day from tens of dollars to $104.25.

Story 8.4: How we forgot to grieve

The discovery of Prozac was one of the greatest contributions not just in psychopharmacology and psychiatry, but in the history of medicine itself. Its strength not only lay in the fact that it could help patients, although that in and of itself would be enough to make its value immeasurable. Where it benefited humanity more is in how it changed two perceptions: the perception of one of the largest groups of illnesses and the perception of one medical specialization. The significance of Prozac cannot be downplayed by the fact that its use gradually blurred and obfuscated the line between depression as an illness and depression as a state of mind or melancholia. Prozac was prescribed even in cases where the goal of treatment was to modify and improve a patient's psychological state rather than to help the patient recover from a depressive disorder. This approach, more common in the U.S. than in Europe, was given the name "cosmetic psychopharmacology" in 1993 by American psychiatrist and journalist Peter D. Kramer in his book *Listening to Prozac*. The term is still valid today.

When Prozac first appeared on the pharmaceutical market in early 1988, nobody had any idea what would become of it. It offered a completely new approach to treating depression. Earlier drugs, like imipramine, acted on the entire spectrum of neurotransmitters, making it a kind of "impure" antidepressant. Prozac targeted one single neurotransmitter, serotonin. Would selectivity overpower non-selectivity in clinical practice? Would "pure" drugs defeat "impure" drugs? If the answer was yes, where would we see the differences the most?

It was not effectiveness in which Prozac outperformed its predecessors; in fact in this respect it was comparable to or even worse than the "impure" drugs. What made Prozac the champion was that it was safer and better tolerated. It had fewer side effects than its predecessors. There were fewer, if any, reports of blurred vision, dry mouth, constipation, sudden drop in blood pressure when standing, and cardiovascular issues, all of which were common with the earlier antidepressants. The milder heart-related side effects made it safer to use and doctors felt better about prescribing Prozac to patients with heart problems and to older patients. The risk of overdose, whether accidental or intentional, was also dramatically reduced. By being safe and better tolerated, Prozac opened doors for itself to psychiatric practices – wide open, in fact.

At first it mainly happened in the U.S., where Prozac gained ground at lightning speed. In 1988, the first year of its use, it was prescribed 2.5 million times – and Eli Lilly made $350 million. Two years later it was eight million prescriptions, and in 2002 it was over 33 million. In the first 20 years of its existence, 40 million Americans used Prozac.

Following the huge success of Prozac, it was not long before new drugs with similar effects in stabilizing serotonin in the synapses began to appear. New molecules were introduced and more pharmaceutical companies joined the crowd.

In 2008, there were 50 manufacturers of antidepressants in the U.S., and most of the drugs being manufactured were in the same drug class as Prozac. In 2013, just 25 years after it was discovered, Prozac and other modern antidepressants had been prescribed over 200 million times. In the 20 years between 1991 and 2011, the consumption of antidepressants in the U.S. more than quadrupled and some 40 percent of Americans had used some kind of antidepressant at least once in their life.

The class of drugs that worked on a principle similar to Prozac was later given the name "selective serotonin reuptake inhibitor," mostly know by the initialism *SSRI*. These were followed by serotonin-norepinephrine reuptake inhibitors, or SNRIs, and noradrenergic and specific serotonergic antidepressants, or NaSSAs.

Europe has always been a bit more conservative in the treatment of depression, so it took a little longer before European psychiatrists began taking advantage of the benefits of Prozac and similar drugs to treat their patients. But when they did, it spread just as rapidly.

Soon, use in Europe was catching up with its use in America. In 2010, nearly 500 million one-month-supply boxes of antidepressants were prescribed. Antidepressant use in Great Britain grew sixfold between 1991 and 2009, and the growth was even higher in the Nordic countries. As the number of boxes grew, so did the profits of the pharmaceutical companies. Take the Danish company Lundbeck, historically one of the most successful companies engaged in the development of drugs for mental disorders. Its operating income in 2004 was 7.5 times larger than in 1994 – it went from 1.3 billion Danish kroner to 9.7 billion.

What is the reason for the huge increase – the largest in the history of pharmacology – in antidepressant use that began in the 1990s and continues more or less to this day?

First of all, it was the properties of the new antidepressants, particularly the favorable side effect profile as already mentioned. Prozac and its successors were appreciably better tolerated and safer than the drugs that came before them. Psychiatrists could prescribe them on an outpatient basis, whereas the major side effects associated with the previous generation of antidepressants meant that they were hesitant to employ pharmacological treatment for outpatients. The new antidepressants were soon being prescribed by more than just psychiatric practices; other specialists were also prescribing them, as were general practitioners. Prozac and the other new antidepressants were also expedient tools in the hands of psychiatrists who leaned towards psychoanalysis, because they made contact with the patient simpler and more effective. As often happens in medicine, the availability of treatment leads to better knowledge of the illness itself, which then leads to an increased number of patients who can benefit from the treatment. Patients with depression suddenly found out that their mental disorder, which their doctors had been struggling to treat, could now be treated a whole lot faster with the new drugs.

But the properties of the drugs alone, no matter how positive, could not have caused such a huge growth in use. There had to be other factors that were not directly associated with the properties of Prozac; factors that reflected changes in the world beyond the drugs themselves.

First of all, there was a change in psychiatry as a medical field. In the late 20th century, the diagnostic criteria for evaluating psychiatric illnesses, depression included, underwent substantial change. Simply put, the seriousness of the illness was viewed differently, which meant treatability was viewed differently as well.

In the 1970s, hospitalization was still typically required for the treatment of severe endogenous depression, and although the first generation of antidepressants was already in use, the most effective method for managing the illness was electroconvulsive therapy. Evaluated this way, depression was an infrequent illness and it affected on average only one in 10,000 people. Less paralyzing and more frequently occurring conditions, such as depressive reaction and depressive neurosis, were not considered to be actual depression, and treatment, if there was any, was more along the lines of psychotherapy. The change in how depression was approached, from the 1980s onward, was that conditions that were less severe than endogenous depression were now being included in the definition of depression. When the new criteria were applied, it was found that it was not one person in 10,000 who suffered from depression: it was one in ten. The new diagnostic criteria also created an entire new group of disorders which had previously not be diagnosed or evaluated separately, such as social anxiety disorder, anxiety and panic attacks, and other disorders that manifested in fear and anxiety. These are not trivial illnesses. As they were more closely studied, it was discovered that a large part of the population suffers from these illnesses. In the course of one year, nearly 20 percent of Americans suffered from agoraphobia – the fear of open or crowded places – and one in 20 suffered from anxiety disorders.

The important thing was that, like depression, these illnesses responded well to drugs that modify the number of neurotransmitters in the synaptic clefts. This meant that, along with Prozac and the drugs that came after it, new disorders were being discovered and a large number of patients suddenly became treatable. Clinical studies began to appear that confirmed the theoretical expectations. Apart from treating depression, antidepressants were now being used to treat disorders that manifested in anxiety, panic attacks, fears, and obsessive-compulsive behavior. The possibilities of Prozac and other new generation antidepressants were practically limitless.

There is a third factor that caused the massive growth in antidepressant use. Pharmaceutical companies got involved. They began to invest heavily into new drug research and they backed clinical studies that confirmed the successful treatment of a growing number of diagnoses. Research investments made by the previously mentioned Lundbeck grew tenfold between 1994 and 2004. Investments were not limited to research and development, with the costs of sales and advertising of the new drugs also on the rise. Large numbers of pharmaceutical representatives were hired and pharmaceutical companies more generously sponsored national and international professional events for a growing number of doctors. And doctors were prescribing more and more antidepressants.

The fourth and final factor cannot really be quantified, or even proven, unlike the first three. It is the social changes that have taken place over the last few decades. One of these changes is that we have somehow forgotten to grieve. Just a few generations ago – and it still applies in Eastern cultures – the surviving family members were entitled to grieve. Society for the most part fully accepted grief and the ways in which it was expressed, so the bereaved were allowed to withdraw for a time to grieve, whether it lasted for six months or five years. It usually helped. But today, few people can afford to withdraw from life and spend time grieving. Society has changed, but our emotional constitution has not.

Now, it is no longer necessary to go through the long and arduous process of coming to terms with grief or emotional reactions to one's own insufficiencies. Just pop a Prozac.

Concluding remarks

Over the years, Prozac has become a true celebrity. The world's most widely read newspapers and magazines gave it plenty of space and the world's biggest news agencies reported on it. Prozac's properties were exalted, but it was also the subject of various lawsuits. It was celebrated, condemned, then celebrated again.

We end this chapter with a few thoughts from Edward Shorter's book, *A History of Psychiatry: From the Era of the Asylum to the Age of Prozac.*

"Inserting Prozac into the history of psychiatry requires untangling good science from scientism. Good science lay behind the discovery of fluoxetine (Prozac) as a much safer and quicker second-generation antidepressant than imipramine and similar drugs. Scientism lay behind converting a whole host of human difficulties into the depression scale, and making all treatable with a wonder drug."

Prozac was hugely beneficial for humanity. It helped mental illness to become acceptable in the eyes of ordinary people. The "lunatics" that were always received with so much horror and fear were transformed into humans that could be helped.

9

Viagra

To tell the story of Viagra[1] and its various predecessors objectively, we must take a look at what it is that Viagra treats. We cannot avoid the issue of male impotence – or erectile dysfunction as it is now known – and erectile dysfunction is a problem of the penis. This male organ therefore takes center stage throughout the stories in this chapter.

Since the dawn of Western culture, the penis has been more than just a part of the male anatomy. It is also the quintessence of the role of a man, a measure of his sexuality and masculinity and consequently, a measure of his place in the world. That men have a penis is a biological fact. How they think about it, how they feel about it, and what role it plays in their lives are things that have little to do with biology. While the organ itself has remained the same, the understanding of its anatomy and opinions on its role in life vary from culture to culture and from one era in history to another.

Nothing else so aptly and succinctly characterizes the role of the penis in a man's life as the number of names given to this organ in the different languages. Technical terms, slang, jargon, vulgar words, emotional and abstract words, personifications. At one point, there were around 100 expressions in Latin and modern languages have roughly the same number of words for penis. In English, the language of Shakespeare, contemporary linguists have counted 160. That is 160 words for one single organ.

[1] Here, we face the same terminological dilemma as we did with Prozac/fluoxetine in Chapter 8. Since the world knows Viagra by that name rather than as sildenafil, the brand name is used in this chapter.

V. Marko, *From Aspirin to Viagra*, Springer Praxis Books,
https://doi.org/10.1007/978-3-030-44286-6_9

Story 9.1: The autodidact of Delft and the penis's status in history

The 17th century was the Baroque Period, the era of trade expansion, French dominance in Europe, the colonization of America, the Thirty Years' War – and the Scientific Revolution. That would be Scientific with a capital "S". By the end of that century, Europe had knowledge of electricity, the law of universal gravitation and movement, the shape of our solar system, air pressure, the logarithm, the telescope, the microscope, and mechanical calculators. Most of these things can be credited to luminaries like Galileo Galilei, Johannes Kepler, René Descartes, Blaise Pascal, Isaac Newton, Gottfried Wilhelm Leibniz, and many, many more. Scientists with a capital "S."

But there was one more man. He was no scientist, not even with a lowercase "s." He was the owner of a small textile shop in Delft in the Dutch Republic, and an occasional chamberlain for city hall in his hometown. He was a simple, self-taught man who never finished his formal education, but he is responsible for one of the greatest inventions in the history of science. His name was Antonie van Leeuwenhoek and he invented the microscope.

He was born Antonie Philips van Leeuwenhoek on October 23, 1632 in Delft. Early on, his private life followed along the same lines as thousands of other common people in those times. His father, a basket maker, died when Antonie was only five years old. His mother remarried, but his stepfather would soon die as well. He was briefly in the care of his uncle, but he was on his own from the age of 16, working as an apprentice in a textile shop in Amsterdam. After earning a little money, he returned to Delft and opened his own textile shop, which gave him financial security for the remainder of his life. He married at the age of 22 and fathered five children, of which only a daughter, Maria, survived. When his wife passed away at a young age, he married a second time.

Gradually, however, his life began to veer further and further away from the life of the average town dweller and Antonie van Leeuwenhoek's social status began to rise in Delft. City hall gave him several appointments. He was chamberlain, land surveyor, and in charge of taxation on wine imports to Delft. He was also the model for the paintings *The Astronomer* and *The Geographer* by the brilliant Delft painter Johannes Vermeer, who was his neighbor.

How did the owner of a textile shop become one of the most revered and famous people of his time; a man who corresponded with one of the greatest scientific institutions – *The Royal Society of London for Improving Natural Knowledge*; a man who was visited by the great scientists as well as by great monarchs, such as Peter the Great of Russia and the Dutch Prince of Orange, William III?

The secret of his success lies in small glass pearls. These pearls were used by textile merchants as magnifying glasses to check the quality of cloth, but Leeuwenhoek was unsatisfied with this simple approach and began thinking about how it could be improved. He affixed a glass pearl to a small brass plate that was welded to a stand. He then placed another plate beneath the one with the glass pearl, to which he attached the object – and the first microscope was born. It was an instrument that simply cannot be compared to today's microscopes: it was tiny, only about five centimeters, and the pearl, which acted as the magnifying glass, was only one millimeter across. The entire instrument had to be held right up against the eye. But it worked perfectly.

Leeuwenhoek made dozens of these single-lens microscopes, nine of which have been preserved. They all have a magnification power of 275. Apart from the way in which he affixed the small magnifying lenses, the secret of this magnification power was in the original method the clever textile merchant had invented for preparing the glass pearls. As if to prove what a smart merchant he was, he never disclosed this secret to anyone and as a result, the quality of his magnifications surpassed those available in all of educated Europe for many years. His method was not replicated until 300 years later, in the late 1950s.

Antonie van Leeuwenhoek soon discovered that his microscope had uses other than determining the quality of cloth, and he began to study natural objects. Day after day, he became more fascinated with the possibilities his invention presented and he became completely obsessed. He let his daughter run the business and turned his full attention to his scientific hobby. Although he was not a trained scientist, he was an astute observer and, in particular, he was a patient and industrious researcher. He studied everything. In 1647, he was the first to describe single-celled organisms and he later described in detail the size and shape of red blood cells. He discovered bacteria in water, and it took 100 years before someone else was able to replicate the discovery.

But he went down in history for making one more groundbreaking discovery. The Delft autodidact found the answer to the questions humanity had been asking since time began. Who are we? How are we created? Where do we come from? It happened in 1677, the year he first described human spermatozoa.

Leeuwenhoek made his discovery public in one of his many letters addressed to the Royal Society. It began with the words: *"What I describe here was not obtained by any sinful contrivance on my part. The observations were made upon the excess with which Nature provided me in my conjugal relations."* This was a necessary opening in order to avoid being accused of masturbation, which was considered a grievous sin by the Church. He then went on to describe what he discovered in his sperm under the microscope. He called his discovery *animalcules*. They were so miniscule that "*...I judge a million of them would not equal in size a large grain of sand. Their bodies which were round... ran to a point behind. They were furnished with a thin tail... They moved forward owing to the motion of their tails like that of a snake or an eel swimming in water...*"

Antonie van Leeuwenhoek's discovery was not just significant in terms of human biology. In a way, it was also a philosophical milestone. Leeuwenhoek had seen something nobody before him ever had. His observation set off a change in the understanding of humans as a species, a change in man's relationship with God, nature, reproduction, sexuality and, naturally, the penis. His discovery started the process of demythologizing the male reproductive organ.

Since the beginnings of Western culture, the penis, a relatively small part of the male physique, has been a distinctive symbol of creativity and creation. This goes back to the ancient Sumerian god Enki, whose semen formed the Euphrates and Tigris Rivers and who used his actual penis to dig the first irrigation canals. The penis had a similar creationary function in the religion of the ancient Egyptians. The creator god Atum needed no one, only himself. "*I created on my own every being. My fist became my spouse.*" The penis as creator and victor also plays a critical role in other Egyptian myths. It is therefore not surprising that the ancient Egyptians saw a connection between potency and triumph on one hand and impotence and defeat on the other.

In ancient Egypt, the triumphant power of the male organ was elevated but the penis itself usually remained concealed beneath a simple short skirt. The ancient Greeks, on the other hand, not only revered the penis, they also flaunted it. In Greek mythology, and in Greek society as a whole, it was a part of the hierarchical system associated with strength, power, male dominance, and the idealization of the nude male body. In the gymnasium, a sort of factory for masculinity, the athletes exercised in the nude, and for unmarried Athenians nudity was a sign of their citizenship. The residents of Athens encountered the naked organ on a daily basis. It was part of the images depicted on vases, pitchers, and other common objects. It was carved into stone columns dedicated to Hermes found at many intersections and corners. Ancient Greeks organized annual processions to celebrate Dionysus, where selected men carried fetishized phalluses. The largest of these, which was gold-plated, was allegedly 180 feet long.

Phallic depictions were part of daily life in Rome as well. They could be found in stone floors, public baths, and the walls of houses. They were supposed to bring good luck, or at least chase away bad luck. Young boys in Rome wore small phallic amulets as a sign of their masculinity.

The fascination with the phallus was personified in one extraordinary Roman god. While the majority of male inhabitants of the Roman pantheon were big and strong, this one, called Priapus, was small and ugly. But he had a huge penis. It was so big that in some of the art he is depicted in, it is the same size as the rest of his body. As with most deities, Priapus, son of Aphrodite and Dionysus, was taken by the Romans from Greek mythology. However, while he was just a minor deity in ancient Greece, the denizens of ancient Rome gave far greater weight to Priapus and his gigantic phallus.

It was said that Aphrodite was very ashamed of her son's unsightly body; but the women of Lampsacus in Asia Minor, where Priapus settled, were of a different opinion. They were extremely satisfied with his proportions. This caused resentment among the male population of Lampsacus, who banished little Priapus from the town. The Gods exacted revenge by bringing down a venereal plague on them and the only way to avert this plague was by respectfully asking Priapus to return. Furthermore, they had to worship him as the god of gardens and livestock, and a macrophallic symbol of his virility was to stand in every home.

The word penis is from the Latin slang word for tail. Curiously, it was not the most commonly used name in Latin. The most frequently used words were *mentula* and the even more vulgar *verpa*, the origins of which are still debated today. No matter what it was called, the penis – especially the large penis – was worshipped and revered by the ancient Romans and they, too, considered it to be an important symbol of strength and power.

The penis gained a completely different status in Christianity. While it was a symbol of potency, masculinity, strength and power in ancient times, Christianity placed it practically at the opposite end of the spectrum. The Son of God was born of a virgin and ascended to the heavens. His asexual creation and life were the antithesis of the sinful existence of man. The soul is pure but the body is wicked and no other organ is quite as wicked as the penis. If we were to look for the date when this historical change took place, we would probably find it somewhere around the late fourth or early fifth century and we would look to Saint Augustine.

As a young man, this future saint did not lead a very virtuous life. He had numerous affairs and he lived in a relationship without the union of marriage for 13 years. The great change in his life occurred when he converted to Christianity at the age of 31 and began studying the Book of Genesis. In it he found the answers to his questions, and they redefined how human sexuality – and thus the penis – would be viewed for the next millennium and a half.

According to Augustine, both the cause and the consequence of Adam's original sin was lust. The symptom of this lust was an erection that a man cannot control. This turned men into slaves to their reproductive organ and made them powerless against it. The penis went from being a symbol of power and virility to a symbol of powerlessness and weakness, and for centuries, it disappeared from public life. Nudity was a sin, not only in daily situations but in art as well. This attitude clung tightly for many long centuries, as the great Michelangelo found out for himself in 1504 when the dregs of Florentine society threw stones at his magnificent statue of David. Thirty years later, when he completed *The Last Judgement* on the altar wall of the Sistine Chapel, Pope Paul IV ordered all the genitals to be painted over.

In spite of all the obstacles, the Renaissance did actually bring about a change in how nudity was viewed, and that allowed the male genitals to be rediscovered in artistic renderings. It was Leonardo da Vinci, Michelangelo's older colleague and a famous artist in his own right, who was particularly fascinated with the penis. Apart from his many famous paintings, he also created more than 5,000 pages of drawings and sketches. These contain designs for flying and submarine machines, military inventions, architectural and urbanism plans, and a very large number of anatomical drawings. Leonardo was obsessed with the architecture of the human body, including the genitals. It is actually from these anatomical sketches that we can see the master's fixation on the penis and the male genitals are the star of many of his drawings. He drew genitals separately, as a part of the male body, and as a party to sexual intercourse. All the drawings carry the master's typical anatomical preciseness, and in fact it was da Vinci's work that took the first step toward the secularization of the penis.

More than 150 years separate the anatomical sketches of the Renaissance genius and the discovery of animalcules by the Delft autodidact. Over that time, the penis went from private drawings to official anatomy atlases and its function was described, as was the process of erection. In addition to being a symbol, whether of power or powerlessness, the penis gradually became an object of science. The role it plays in reproduction was still understood only on the level of a symbolic act. It took one Antonie van Leeuwenhoek to show everyone the product of the erect member, the product responsible for the creation of a human being.

Leeuwenhoek's animalcules were not the sperm that we know today. He and many who came after him believed that a complete human was concealed inside the sperm – only smaller. The female loin was viewed as simply an incubator of sorts, in which a very small human, a *homunculus*, slowly grew and developed. This was the prevailing hypothesis for some time, even though it could not answer the question of how it was possible that typically only one out of the astronomical number of homunculi would make itself at home and develop inside a woman's body.

Maybe in the case of Antonie van Leeuwenhoek's discovery, as with all discoveries, the true contribution is that it raised more questions than there were before.

Story 9.2: The son of a Russian vodka maker and elixirs of youth

If we were to cast aside political correctness to start off this story, we would claim that men mainly want to be strong and women mainly want to be beautiful. What both genders probably have in common is that we all want to remain youthful as long as possible. That is because – continuing with the political incorrectness – for men it would mean remaining strong and for women it would mean remaining beautiful. For the "weaker" sex (more political incorrectness), the claim ends there; for the "stronger" sex, it goes one step further. A man's youth (and strength) is closely tied to the very important ability to have an erection, at any time one is needed.

Various elixirs of youth have therefore accompanied humankind throughout our cultural history, right alongside fountains of youth and the philosopher's stone. The first medical document to describe anti-aging treatments is an Egyptian papyrus from roughly 1600 BCE. It contains a spell that turned an old man into a 20-year-old youth. In ancient Greece, a plant known as satyrion was a popular rejuvenating treatment. The name of the plant is derived from the satyrs, the companions of Dionysus who were known for having a permanent erection. The plant was so popular that it was nearly wiped out by the ancient Greeks, but it managed to hang on and today it grows almost everywhere around the globe and is known by the official Latin name *jacobea vulgaris*. Most rejuvenation preparations were also aphrodisiacs, in confirmation of the claim made in the previous paragraph. We will take a look at them in the next story, but for now let us briefly look at another method of rejuvenation that was popular in England around the year 1000. It was known as love bread. Nude young women would run around in a circle in a wheat field, and the wheat that was crushed beneath their bare feet was used to bake bread. It just goes to show that probably the most reliable aphrodisiac is the imagination.

But enough of these old and unscientific methods. The 20th century saw the introduction of completely new methods of rejuvenation; modern and scientific methods based on the latest medical knowledge. The first swallow was a discovery made in the late 19th century by Charles-Édouard Brown-Séquard, the multitalented scientist born in Mauritius to a naval captain and a Creole mother who had his own story in the chapter about insulin. His story is worth repeating here because of the announcement he made to members of the *Société de Biologie* in Paris on June 1, 1889. This is a day that can be considered a milestone, the day on which the elixir of youth shifted from the realm of spells and superstitions to become a part of organized medicine. Already 72 years old at the time, Dr. Charles-Édouard Brown-Séquard presented to the esteemed members of this biological society a report on an experiment he had tried out on himself. He had been subcutaneously injecting himself daily over a two-week period with fluid extracted from the testicles of dogs and guinea pigs. The results were extraordinary. There was a significant improvement in his strength and stamina, so that he could lift heavier loads and walk faster up the stairs. Additionally, his urine stream was now 25 percent more powerful than before.

The medical community was nonplussed. In one obituary, the last years of his research were characterized as "*outright errors of senility*," but the general population welcomed his new method with open arms. The press ran with the news – and pharmaceutical manufacturers paid attention. Newspaper headlines and magazine articles quickly appeared, calling it the elixir of youth and a panacea. The popularity of Charles-Édouard Brown-Séquard rose to dizzying heights and his portraits and caricatures appeared on the front pages of newspapers and magazines on both sides of the Atlantic. While the newspapers published articles, jokes, and odes, the pharmaceutical manufacturers prepared and sold countless quantities of animal extracts with names that included various forms of the word *elixir*. In the early 20th century an English company, C. Richter & Co., sold an extract from the testicles of rams and bulls as a product using the name of the discoverer – Séquardine. Over time, however, there was mounting evidence that if anything about the concoctions made from testicle fluids was working, it was purely a placebo effect: the hope and belief that lost youth and strength could be reclaimed.

On the other hand, what if Brown-Séguard's method was not being reproduced successfully because there was too little testicle extract in the elixir? If that was the case, then would it not help to transplant the entire organs of healthy and strong animals into patients? Goats, for example, are known for their sexual prowess, or maybe rams, or the closest animal to humans, the chimpanzee. That brings us to the main character in this story: Serge Voronoff, who was born Sergei Abramovich Voronov.

Sergei Abramovich was born into a Jewish family in the town of Shekhman in the Tambov Governorate, Russia. His exact date of birth is unknown, but the date of his circumcision was July 10, 1866. His father, Abram Veniaminovich Voronov, was a local vodka distiller. He must have made good vodka, because Sergei was able to go to Paris at the age of 18 to study medicine. Sergei became a French citizen at the age of 29 and changed his name to become more French-sounding. In 1896 he went to Egypt, where he worked as a doctor in the court of the ruler Abbas II. He met and later married Marguerite-Louise Barbe, whose father was Ferdinand de Lesseps, the famed developer of the Suez Canal.

Serge Voronoff's stay in Cairo triggered his interest in glands and rejuvenation, to which he devoted the rest of his life. It all started with his observations of the eunuchs in the monarch's court. They were mostly fat and sickly men, whose hair turned white at a young age and who rarely lived to an old age. Voronoff thought to himself, what if these changes are caused by the removal of their testicles? What if the process were reversible, and youth, strength and longevity could be achieved by adding back the absent male organs? After 14 years in Egypt, he returned to Paris in 1910 and began his great and successful career. It would be a while before it would take off, but after the end of World War I there was nothing to stand in his way.

His first wife died shortly after they returned to Paris and some authors allege that she died during secret alchemy experiments conducted by a young occult artist. Her death allowed Serge Voronoff to meet and later marry Evelyn Bostwick, the daughter of a wealthy oil magnate. She was a nurse in the Boer War, holder of the Military Cross for bravery, and a skilled laboratory assistant. She was also alcohol and drug dependent and Voronoff was her fourth partner. They were married in 1920, but the marriage only lasted just over a year, as Evelyn died suddenly at the age of 48, probably from a combination of drugs and alcohol. However, her daughter from her first relationship claimed that her mother was killed by Serge Voronoff.

Whether that is true or not, the inherited wealth gave Voronoff complete financial independence and allowed him to dedicate himself to his experiments. He first attempted his rejuvenation procedure on animals, starting with an old ram. He grafted the testicles from a younger animal onto the older animal and observed. He later wrote that the sheep's wool thickened and its sex drive increased. From 1917 to 1926, Voronoff performed around 500 transplantations of the testicles from younger male sheep, goats and bulls to older ones. He believed his operations were successful because this new strength could absolutely be observed in the older males.

If it worked on animals, why not try it on humans? Since primates are man's closest relative, Voronoff decided to use the testicles of chimpanzees. He expected he would be able to turn back a man's clock by 20–30 years.

Serge Voronoff performed the first official transplantation of monkey testicles to a human on Sunday, June 12, 1920, by implanting a thin slice (a few millimeters) of monkey testicles into his patient. The thinness was to ensure it would better fuse with the recipient's tissue. His work caused a sensation and from that moment, the surgeon became famous not only in Paris and Europe, but across the Atlantic as well. There was a huge waiting list for the operation. Three clinics were opened in Paris, all at very prestigious addresses, and a fourth clinic was set up in Algiers. Hundreds of men in France, and thousands around the world, underwent the operation. Male chimpanzees were fairly expensive and the demand exceeded the supply from Africa, so Voronoff decided to start his own monkey farm. When the French authorities prohibited the farm, he moved it to Italy, just across the French border, into the garden of the posh manor house known as Villa Grimaldi. The manor house still exists as Villa Grimaldi Voronoff, where you can book a room if you are ever in town.

In contrast to what happened with Charles-Édouard Brown-Séquard, Serge Voronoff was celebrated by both the general public and the medical community. In 1923, he was applauded by 700 of the world's preeminent surgeons in London at the International Congress of Surgeons, and he presented his method on the American continent as well. Experts accepted his belief that rejuvenation by grafting could not only positively affect sex drive, but it could also improve memory, enable people to work longer hours, and prolong life. The first elixir of youth, scientifically proven and medically verified, was born.

Its discoverer became a true celebrity, his name splashed across the headlines of newspapers and magazines. He was the subject of serious contemplations and ridicule alike. The expression *monkey glands* became a synonym for rejuvenation, and not just in surgery. According to newspaper articles of the time, New York's subway system needed monkey glands for rejuvenation, while 30 years later, the British Conservative Party needed the same – figuratively speaking, of course. There was even a cocktail named the Monkey Gland, created by Harry MacElhone, owner of Harry's New York Bar in Paris, France.

Serge Voronoff was not a lone wolf in discovering new, medically supported methods of rejuvenation. His most important pack members – and rivals – were in the United States.

First, let us meet Dr. Leo Stanley. He found a solution to the problem that was causing his pack member trouble: although monkey glands were similar to human glands, they were still of animal origin. Ideally, there should be human donors, but where could you find young, strong men who would be willing to give up their masculinity to benefit the masculinity of another man, and an old man at that? Dr. Leo Stanley had one big advantage in that he was the chief surgeon at San Quentin State Prison. He oversaw the execution of three or four young convicts every year, which gave him access to "material" even without the consent of the donors. He began his experiments in 1918 and the first donor was a young African American man, Fred Miller, who was sentenced to death by hanging for first-degree murder. His testicles were donated to another inmate, 72-year-old Mark Williams, who at that time suffered from advanced senility. The old man's condition was visibly improved after the operation, as though he were a younger man, who was even able to get the punchline of jokes. Over the years, Dr. Stanley successfully performed the operation on 643 prisoners and 13 doctors. The recipients boasted of better vision and appetite, increased energy, and a loss of tiredness. The results were downplayed somewhat because all the improvements were self-reported by the participants and therefore subjective, but Stanley himself believed that the transplantation of the testicles of healthy young men into older recipients could improve neurasthenia, melancholy, asthma, and acne. He had some doubts, however, about the effect on tuberculosis.

In any event, Dr. Leo Stanley was not the main player in rejuvenation on the American continent. If we only take into account actual doctors and not quacks, that would have been Frank Lydston. He described his first experience with a testicle transplant in 1915, several years before Serge Voronoff performed a similar operation in Paris. Lydston was no rookie or recent graduate at the time, but a full professor of practical surgery and surgical pathology at the University of Illinois. Over time, he had authored several publications discussing the potential use of the method for other problems. These included improving blood pressure, the treatment of arteriosclerosis, slowing down senility, the treatment of various psychopathologies and, naturally, prolonging life. As a true experimenter who believed in his discovery, Lydston also performed a few operations on himself. He once surprised one of his colleagues by dropping his pants in front of him to show the result of one of those operations. The astonished colleague saw three testicles dangling below Professor Lydston's penis: his own two, and one that had been donated by an executed murderer and implanted. Three is greater than two, and so the professor not only had an increased sex drive, but his thinking was faster and clearer. Furthermore, and no less important, the implantation of the testicles of a criminal did not stimulate any criminal urges within him.

Finally, we have one more method that differs slightly from the preceding ones. It was invented by Viennese physician Dr. Eugen Steinach, professor and director of the Institute for Experimental Biology of the Academy of Sciences in Vienna (*Biologischen Versuchsanstalt der Akademie der Wissenschaften*). His method is called the vasectomy, a procedure that prevents sperm from entering the urethra from the testicles. It is a commonly used form of male sterilization and is essentially one of the most effective contraceptive methods available. Dr. Steinach believed that by cutting and tying off the vas deferens, sperm would return back into the body where they would have a rejuvenating effect. Additionally, in older men it would increase hormone production in the testes, possibly attaining the same level as when they were young and strong. Like his rivals, he offered conclusive evidence of the effectiveness of his method, including hair growth, improved vision, and therapy for various diseases. His method was much simpler in comparison with the others mentioned above. It became so popular that instead of using the long phrase "*getting a vasectomy for rejuvenation purposes*," people just said "*get Steinached*." Probably the most famous man to get "Steinached" was well-known Vienna resident Sigmund Freud. After the operation, he announced that he was extremely satisfied. Eugen Steinach himself was nominated six times for the Nobel Prize.

This is where we conclude the "*medically proven elixir of youth*." The great enthusiasm did not last long, only about a quarter of a century. Increasing evidence in the late 1920s and early 1930s showed that if there had ever been any rejuvenating effect, it was very fleeting. This news was encouraging to those who still believed in it enough to undergo rejuvenation procedures, but there were fewer and fewer of them. The discovery of testosterone in the mid-1930s brought a definite end to testicle grafting. During their lifetimes, the names of the rejuvenation protagonists were frequently spoken with the same respect as the names of the somewhat older greats like Louis Pasteur and Robert Koch, but they have since been long forgotten. Their portraits do not hang in doctors' offices and their names cannot be found in medical textbooks. Actually, they sank into oblivion even before they passed away and while you might probably have expected it, not one of them significantly prolonged their own lives. Serge Voronoff died at 85, Eugen Steinach was 83, and Frank Lydston, the man with three testicles, was only 65 years old when he died. When Serge Voronoff died, he was depressed, disappointed, and forsaken. On the upside, he had spent years enjoying the massive wealth that monkey glands brought him.

We now know that despite their colossal, often fanatical confidence in their own methods, they were methods that simply could not work. If there is one thing that transplantation medicine still has trouble with, it is immunity, and immunity is what rejects foreign tissue. The transplanted testicles or sections of testicles had no way to fuse with the existing tissue and remained, shriveled up, inside a man's testicles as evidence of his efforts to achieve an impossible dream.

But questions also remain that must be answered. How is it possible that for 25 years, millions of people around the world believed in the rejuvenating effects of implanted animal testicles? How is it possible that everyone who underwent the operation (the vast majority, anyway) claimed that the animal glands had an indisputable rejuvenating effect? It was not only the patients – the elixir of youth was supported by a large portion of the medical community and with clear theoretical substantiation that the entire procedure was legitimate.

The entire affair of seeking, and finding, an elixir of youth through the grafting of animal glands is an extreme example of the placebo effect: a substance or procedure that objectively has no specific action on the disease or condition it is used to treat. Although the placebo is ineffective, its administration encourages expectations to which the body responds with changes that are then attributed to the substance or procedure.

A piece of dead tissue could not have had any rejuvenating effect, but it carried with it the expectation of newly-rediscovered youth and of having the sexual prowess of a 20-year-old. The expectation, the hope, was so powerful that patients had no trouble fooling themselves, others, and their doctor that it worked. In many cases, it probably did work for a while, but it was no elixir of youth.

It seems that the last men to undergo monkey gland treatment were the football players of England's First Division team Wolverhampton Wanderers in 1937. According to the website *These Football Times*, after they took the treatment, they beat Everton 7–0 and Leicester City 10–1.

But at least one good thing came out of all these rejuvenation efforts: the Monkey Gland cocktail. So here is the recipe for the cocktail, as published in the *Savoy Cocktail Book* in 1937:

2/3 dry gin
1/3 orange juice
3 dashes Absinthe
3 dashes grenadine

Cheers!

Story 9.3: The biggest charlatan and the deepest desires of men

While the protagonists of the preceding story – Serge Voronoff, Leo Stanley, Frank Lydston, and Eugen Steinach – were all well-known and respected physicians, Dr. John Romulus Brinkley, M.D., Ph.D., M.C., LL.D., D.P.H., Sc.D., was a common grifter and charlatan. Well, an exceptional grifter and a colossal charlatan. The most colossal of all charlatans. This is his story.

He was born on July 8, 1885 in North Carolina into the family of impoverished mountain man John Richard Brinkley. The story of how John Romulus was born is complicated. The first time his father married, he had been underage and the marriage had to be annulled. After he became an adult, he was married four more times and he outlived every single one of his wives. He married Sarah T. Mingus at the age of 42, and her niece Sarah Burnett came to live with them. The younger of the two Sarahs was the mother of John Romulus. The women around John Richard all seemed to die young, and that was the case with Sarah, who died when her son was five years old.

John Romulus attended the local one-room log cabin school, where classes were usually held in the winter months. He learned to use a telegraph and spent a few years working for various railroad companies, but the boring life of a telegraph operator was not for him. He wanted to save humanity, free the slaves, and cure incurable diseases. He dedicated his life to the last of those goals – except that he was not curing incurable diseases. In fact, he was not curing much of anything.

He began by joining the large ranks of traveling medicine shows that went from town to town and village to village selling miracle cure-alls. Traveling medicine shows flourished in the United States in the late 19th and early 20th centuries. Hundreds of so-called patent medicines were widely available and they offered a cure for practically everything. Think *Dr. Ryan's Incomparable Worm Destroying Sugar Plums*; *Bardwell's Aromatic Lozenges of Steel*; and *Lydia Pinkham's Vegetable Compound* for all existing women's problems. There were over 40 various tinctures for diabetes alone. The main ingredient in most of these was alcohol, the content of which could be quite high. As for the other ingredients, they were ineffective or worse. *Paine's Celery Compound* had an alcohol content of 21 percent, and *Hostetter's Stomach Bitters* had 44 percent. It is hardly surprising that these patent medicines were so popular.

In the meantime, John Romulus married his childhood friend, Sally Margaret Wike, and together with another young couple they traveled the eastern United States with their medicine show. The show consisted of some short dancing and singing performances, after which the main show would begin, involving a main character who was suffering and dying because they could not get the help they so desperately needed. Then John Romulus Brinkley himself would take the stage, dressed in a cutaway coat and hat, praising the benefits of his cure-all tonic. Business was good, but after a time the crew split up and then the Brinkleys welcomed a daughter to their family. It was time to settle down. John Romulus went back to his original profession as a telegraph operator – and began attending medical school. His alma mater was just one of the many unaccredited medical institutions in Chicago, but he took his studies seriously, nonetheless. He went to school during the day and earned a living at night, which did not leave him a lot of time for family, especially since he usually spent whatever free time he did have in between work and school in a bar. Sally was none too happy about the situation, and one day she took their daughter and left John Romulus. She soon realized, however, that her husband had no money to pay her any support so she went back to him, only to run away again shortly thereafter. There was nothing left for Brinkley to do but drop out of school after three years, leave Chicago without paying his tuition, and follow his family. At that point, they already had two daughters and were expecting a third. He could not continue his studies because his previous school would not provide his transcripts since he owed them money.

That brings us to the end of part one of the life of John Romulus Brinkley. We catch up with him a few years later, in 1913 to be precise, in Greenville, South Carolina. Meanwhile, patent medicines had come under the strict control of federal law, ending their golden age. Something more modern would have to be found, and what was more modern in the early 20th century than electricity? Brinkley, now without a family but with a new partner, James Crawford, opened a practice with the proud name of *Greenville Electro Medic Doctors*. They took out ads in the *Greenville Daily News* asking men if they were manly and full of vigor. As expected, many thought they were not. The partners split up their duties in the office. While Crawford manned the reception desk and welcomed clients, made diagnoses, and sent patients to the treatment room, Brinkley manned the treatment room and injected colored water into the behinds of resolute men. If anyone asked what was in the injection, the answer was that it was electric medicine from Germany. All this cost just $25, or more than $600 in today's terms.

As it was strictly a sham, the business did not last very long. They both knew it and they fled Greenville by summer's end, leaving behind a large number of bounced checks. Crawford knew some girls in Memphis, and after spending some time in Mississippi, the partners moved on westward to Memphis. It was late summer and both young men had plenty of money from their previous gig, so there was fun to be had in the company of young ladies. One of the young ladies was Minerva Telitha Jones, the 21-year-old daughter of local physician Tiberius Gracchus Jones. She and Brinkley hit it off right away, so much so that they were married four days later. Not wanting to spoil the wedding mood, the groom failed to mention to anyone that he was still married and the father of three girls; that would come out later. The newlyweds did not enjoy their honeymoon for long, as the Greenville sheriff showed up in Memphis with a warrant and both men were arrested for forgery and practicing medicine without a license. When the lawyers added up their debts to the local merchants, it worked out to be in the thousands of dollars. The bulk of the debt was paid by James Crawford while Brinkley's part was covered by his father-in-law.

John Romulus Brinkley fell off the radar for some time, but resurfaced in October 1917 in Milford, Kansas. He had been busy divorcing Sally so that he was able to marry Minnie, this time legally. He served for two months and 13 days in the U.S. Army as a medical officer, and he "completed" his medical training at the *Eclectic Medical University* (EME) in Kansas. While the EME was a diploma mill where diplomas were handed out to those who did not actually attend classes, the diploma itself looked dignified and allowed Brinkley to practice medicine in eight states.

Milford had a population of around 2,000 when the Brinkleys arrived. The town was looking for a doctor, but it had no paved roads or sidewalks, or even a water supply or sewage system; in fact, it did not even have electricity. In early October 1917, none of the residents could have guessed that their town would become famous throughout the entire United States, or actually, the entire world. It was difficult for the Brinkleys in the beginning. They earned some money from tonsillectomies and other similar minor surgeries, but gradually things improved. Brinkley would likely have ended up merely as a fairly prosperous country doctor except that one day, not two weeks after he had opened his practice in Milford, a local farmer came to his office. His name was William "Bill" Stittsworth and this visit completely changed the life of John Brinkley, along with hundreds and thousands of men across the United States of America. It made John R. Brinkley an extremely wealthy man and the protagonist of this story.

It took the 56-year-old farmer some time to work up the courage to reveal the reason for his visit to the doctor's office. Finally, though, he admitted: "*I'm a flat tire*." Stittsworth had tried everything, but to no avail. He was unhappy, his wife was unhappy, and it seemed he would remain impotent for the rest of his life. "*Too bad I don't have billy-goat nuts*," he said. And that is how it all began.

There are two versions of what happened next. According to Brinkley, it was the farmer's idea and although as a doctor he balked at first, he let himself be talked into doing the experiment. The Stittsworths, on the other hand, claimed it had been Brinkley who offered the farmer a large amount of money to undergo the operation. Whatever the truth was, they both reached an agreement. Neither of them wanted any publicity, so the operation was performed two nights later while the town of Milford slept. The farmer brought his goat and Brinkley removed the goat's testicles and implanted them into the farmer. The whole thing took less than 15 minutes.

Although both men tried to keep everything under wraps, it was difficult to try to hide an attempt at curing impotency, especially after Stittsworth inevitably bragged about the success of the operation. Another hopeful came knocking on John Brinkley's door and soon there were more and more people interested in goat testicles. Mrs. Stittsworth was one of them. She figured that if goat testicles were helping the men, then female goat ovaries could help women just as well. Brinkley agreed. A year later, the "rejuvenated" Stittsworths welcomed little Billy Jr. to the family and the incredible career of one tremendous quack took off.

Within a year, in August 1918, John Brinkley opened a clinic in Milford for 16 patients, furnished in the style of a swanky bed and breakfast. It boasted the name *Brinkley Institute of Health*. The patients' rooms were on the first two floors of the three-story building, with the nurses housed on the third floor and the goat cages located behind the building. The owner of the institute also had to deal with the delicate matter of what breed of goat would donate the testicles. Goats are known not only for their sex drive but for their typical stench as well. Goats stink, but one breed, the Toggenburg, is the exception to this rule and so, completely by accident, the Toggenburg goat went down in history.

John R. Brinkley began racking up the successes one after another. Older men bragged about getting back their strength and virility (meaning sexual vigor), and there were so many people waiting for an operation in Brinkley's institute that they pitched tents in the pastures around Milford. When the wives began clamoring for the same rights as their husbands, Brinkley set up a little business on the side, implanting goat ovaries into willing women just as he had done for Mrs. Stittsworth. He promoted the operation as a way to increase fertility, remove wrinkles, and increase breast size.

That was not the end of the good news. After the implantation of goat testicles helped cure a farmer's nephew of a severe mental disorder, Brinkley could add that to the list of applications for his treatment. Apart from curing sexual problems and dementia praecox (now known as schizophrenia), 25 additional diseases and ailments including emphysema and flatulence gradually joined the list of afflictions that could be cured with goat testicles. The main clientele still consisted of impotent men and they were the target audience of his slogans: "*All energy is sex energy*" and "*A man is as old as his glands*." The number of grateful patients grew along with the number of operations he performed and the carousel began to spin faster and faster. The placebo worked wonders. Men were – and still are – willing to believe anything that promises to transform them into "*the ram that am with every lamb*," as another of John's slogans put it. He and his team performed 50 operations a week at $750 each. That equates to about $10,000 each in today's terms. Brinkley even had a special offer for customers with fatter wallets. Inspired by his more esteemed colleagues Dr. Leo Stanley and Dr. Frank Lydston (both of whom we met in the previous story), he started to offer the testicles of young men for transplantation. Naturally, the price was a little higher. Men's testicles were not as widely available as Toggenburg goat testicles and the price for this service was between $5,000 and $10,000. At this pace, the quality of the operation could not be sufficiently guaranteed and every now and then a patient would die, but these were just minor blemishes on the image of John R. Brinkley.

The time had come when he had outgrown Milford. The time had come for him to go out into the big world.

His first trip beyond the borders of Milford was in 1920, to the Park Avenue Hospital in Chicago. He performed 34 operations there, 31 of which were goat testicle implantations in men and the other three were goat ovary transplants in women. Everything was done in the presence of reporters from the Chicago newspapers and magazines. The patients included a judge, an alderman, and a member of high society. The operations were successful and the headlines the next day were full of John R. Brinkley, who could restore the youth and revive the vigor of aged patients. Thanks to the guru from Kansas, "*Chicagoans had found the lost fountain of youth*." He was awarded an honorary doctorate by the Chicago Law School, while the 72-year-old chancellor of the university was one of the successful patients of the "Milford Messiah."

A similar scene played out two years later in Los Angeles, where Brinkley was invited by Harry Chandler, publisher of the *Los Angeles Times* and one of the wealthiest men in California. He performed the surgery on the managing editor of Chandler's newspaper, on Chandler himself, on a circuit court judge, and on an unnamed Hollywood movie star, all successfully. The *Los Angeles Times* featured the headlines: "*New Life In Glands – Dr. Brinkley's Patients Here Show Improvement... Twelve Hundred Operations Are All Successful.*" Brinkley got $40,000, Chandler had a topic for several issues, and Buster Keaton used goat gags in one of his comedy shorts. Just like Chicago, Los Angeles lay at Brinkley's feet.

But not everyone shared the enthusiasm of Brinkley's patients and not all of his efforts ended in victory. For instance, he wanted to open a hospital in Chicago similar to the one in Milford, based on the success of his operations. The local authorities were against it. For one thing, keeping 15 goats on hospital grounds would have been illegal, for another, Dr. Brinkley had no license to practice medicine in the state of Illinois or, by extension, in Chicago. Furthermore, he did not have the education required to obtain a license to practice medicine, as his only doctorate was an honorary one from a law school. The situation was similar when he attempted to open a hospital in Los Angeles. He was rejected, so he returned to Milford.

Milford loved him and why would they not? They thought he was eccentric, of course, but he also shared part of his wealth with the town. In addition to a new and larger hospital, Brinkley paid for new sidewalks and a sewer system, got a new post office and a new bank built, had electricity brought in, and his money paid for paving the road from the railroad station to the town center. He sponsored the local Little League team called the *Brinkley Goats*. He also bought a bear as the first animal for the local zoo.

Beyond Milford, however, Brinkley's fame was slowly waning. He would have to come up with something new to promote his ability to make aged men "*the ram that am with every lamb*" far beyond the Kansas state line. That something had actually been around for some time, but no one had found a way to take advantage of it yet – radio. John R. Brinkley was one of the first to make this medium work to his benefit. He was a pioneer who immediately saw the huge potential of radio broadcasting as a way to spread advertising. He installed a transmitter outside Milford, along with two 300-foot towers, and named the radio station KFKB (Kansas First Kansas Best). He could now begin broadcasting. The program was a mix of country, gospel, and Hawaiian music, horoscope readings, and French lessons. The main feature of the daily program was his own show. Brinkley would talk for hours, usually promoting goat gland transplantation. He was a master at making up advertising slogans and non-existent recommendations and he also had a team of people who thought up slogans and recommendations for him. Here is one that says it all: "*Note the difference between the stallion and the gelding. The former stands erect, neck arched, mane flowing... seeking the female, while the gelding stands around half-asleep, cowardly... Men, don't let this happen to you.*" It was far-fetched, but it worked.

Brinkley invented a program he called the *Medical Question Box*. It was a format that later became popular at numerous radio stations. Listeners would send in their questions about a medical problem, and he would recommend suitable preparations for treatment. The preparations were sold with huge markups and only in a select network of pharmacies that were part of the Brinkley Pharmaceutical Association. Some estimates say that Brinkley's share of the sales amounted to more than ten million in today's dollars.

The people believed in John R. Brinkley, but the official institutions and the scientific community were not so enamored. In 1923, the diploma mill fraud at the Eclectic Medical University of Kansas City was exposed and John R. Brinkley's name was among the 167 "physicians" revealed to be practicing medicine under a license obtained with this medical degree. He was also named in 19 indictments by a grand jury in San Francisco for substantially benefitting from a fake medical degree. A warrant was issued for his arrest, but when it was time to extradite the defendant from Kansas to California, the governor of Kansas refused to give him up, saying that "*We people in Kansas get fat on his medicine. We're going to keep him here so long as he lives.*" Brinkley could relax for a moment and cleverly turned the tables to his own benefit. He went on the radio to say that the "*persecution (of him) was... no more justified than the persecution of Christ.*" But the California arrest warrant was by no means the end of his problems.

John R. Brinkley was not the only quack in the United States, but he was definitely one of those who made the most money off the gullibility of people. He was a constant thorn in the side of organized medicine, represented mainly by the American Medical Association (AMA), in whose journals the quackery of Brinkley and his colleagues was continuously exposed. But, as one subscriber to the *Journal of the American Medical Association (JAMA)* wrote, "*...we who already know, read it, and [those] who should be warned do not subscribe to the JAMA.*"

But the tide turned. In 1930, the Kansas State Medical Board revoked John R. Brinkley's medical license because he "*performed an organized charlatanism.*" Six months later, the Federal Radio Commission refused to renew his broadcasting license, stating that station KFKB primarily broadcast advertisements, its content was obscene, and that Brinkley's invention, Medical Question Box, was not in the public interest. That should have been the end of John R. Brinkley. It was not.

Since he could not perform transplant surgery, Brinkley transferred the clinic in Milford to his assistants. Since he could not broadcast, he sold KFKB. He moved to Del Rio, Texas, a town separated from Mexico by a bridge, then bought a radio license in Mexico and put up a giant transmitter, the largest in the world. It was so powerful that in good weather it reached the entire United States and could even be heard in Canada. The signal was so strong it turned off the headlights of cars and interfered with telephone calls. Local farmers claimed they did not even need a receiver to listen to Brinkley's program; they could catch the vibrations on the metal fillings in their teeth.

John R. Brinkley needed a powerful transmitter. He had decided to run for governor of Kansas, if for no other reason than to be able, as governor, to appoint his own medical board that would then reinstate his medical license so he could start performing operations again at his clinic in Milford.

He lost. It was close, but he lost. He would have won the election had the election committee not disqualified some of the ballots for errors. He ran a second time and again was beaten, just barely, by his opponent.

The fall of John R. Brinkley had begun. A few years later, his archenemy Morris Fishbein, secretary of the AMA, recounted in detail the substance and ramifications of the fraud perpetrated by the "Milford Messiah." Brinkley sued Fishbein, but he lost the case. The jury found that Brinkley was "*…a charlatan and quack in the ordinary, well-understood meaning of those words.*" The judgment opened the floodgates and numerous other lawsuits followed, costing Brinkley a large portion of his wealth. The IRS sensed there was tax evasion afoot and joined in. Mexico cancelled his radio station and Brinkley declared bankruptcy. He died of heart failure in San Antonio on May 26, 1942.

That was the life of John R. Brinkley in a nutshell. There is not enough room here to talk about his travels across China and Japan, where he convinced the Japanese government to make goat gland transplantation "*…compulsory in Japan… in order to rejuvenate aged charity patients*" as one headline read. Also omitted is his trip to Europe, where he tried to obtain honorary degrees after his medical degree had been revoked in America. The only place that would oblige was the university in Pavia, Italy, but even that was later rescinded – by Benito Mussolini himself – under pressure from American medical circles.

It is true that John R. Brinkley was a charlatan, liar, and a quack, but even his biggest adversaries had to admit that he had turned his life into one giant adventure.

It is also true that neither Brinkley, nor the physicians from the previous story, would ever have been as rich and famous had it not been for men's willingness to do anything and everything for their erection – including having pieces of goat transplanted into their scrotum. But the Toggenburg goat gland implants were far from the end of the story. Rejuvenation clinics still exist and they still make considerable amounts of money. Every day, the author's spam folder includes emails with recipes guaranteed to enlarge the penis and there are countless websites offering all kinds of guaranteed products to help you achieve the virility of a 20-year-old, or whatever else you're looking for.

Perhaps even more surprising than the desire to be "*the ram that am with every lamb*" is the continued willingness to believe it can be achieved, or that it has actually been achieved. The placebo effect that improves male potency, at least for a time, is definitely one of the most powerful of the phenomena.

There is a reason why modern sexologists say that male impotence is more often a problem of the head than of the testes.

Story 9.4: A urologist drops his pants and what men are willing to endure

Las Vegas, Nevada has been witness to many things since it was founded in 1905. Over the years, it has become used to plenty of things too. There have been over 80 movies filmed here. David Copperfield, Tom Jones, Elvis Presley, Frank Sinatra, and even Czech entertainer Karel Gott have all performed here. You can see an 84-year-old dancer's show, take a picture on a (non-working) electric chair, or order a cocktail for $1,000 that comes in a toilet bowl-shaped glass. But what happened at 7:00 pm on the evening of Monday, May 18, 1983, was unusual even for Sin City. That evening, a man in sweatpants stood up in front of around 100 formally dressed people, dropped his pants, and showed everyone his erect penis. Nobody escorted him out of the room and nobody called the police. On the contrary, his actions are considered a milestone in the history of one serious human activity.

Impotence was a male problem long before it got its name in the 17th century, and long before it became a medical diagnosis. Loss of virility, lack of courage or desire, weak loins, or just plain old performance anxiety, has been plaguing men since the dawn of time. Records of treatments for impotence are as old as humankind. The list of everything men have been willing to eat, drink or endure just to be able to copulate makes for some pretty remarkable reading, but it is also absolute proof of what a huge problem erectile dysfunction was and is.

Take this recipe found on a 3,700-year-old ancient Egyptian papyrus: "*One part leaves of Christ's-thorn; one part leaves of acacia; one part honey. Grind the leaves in this honey and apply [to penis] as bandage.*" A 3,000-year-old Assyrian recipe recommended eating dried lizards and Spanish flies. The dried lizards ceased to exist as a recipe right along with the Assyrian Empire, but Spanish fly is still around because it often causes male genitals to turn red, which could be mistaken for the desired effect, if one is optimistic. Another Assyrian recipe was far more effective. It called for a woman to rub a mixture of oil and powdered iron on the penis of an impotent man while chanting "*let this horse make love to me*." The combination of a woman's stroke and flattery is probably one of the most effective recipes available. The magical approach that was once used in the prevention of impotence can also be seen in the ancient custom of the groom urinating through the bride's wedding ring during the wedding ceremony. A wonderful example of this is the unconventional painting by the Italian painter Lorenzo Lotto, *Venus and Cupid* (1540).

Most of the ancient, as well as the not-so-ancient recipes recommended eating parts of animals that were known for their sexual potency, or the parts that were shaped like what they were trying to achieve. The former group includes goat testicles (remember goats from the previous story) cooked in milk and butter, and the genitals of roosters and horses. The truly courageous could have hippopotamus testicles. During orgies in Nero's times, they would serve a beverage of crushed goat (there we go again) and wolf testicles. The wolf can also be found in a recipe from the 13th century: "*If a wolf's penis is roasted in an oven, cut into small pieces, and a small portion of this is chewed, the consumer will experience an immediate yen for sexual intercourse*."

The latter group of recipes that called for eating animals and animal parts that are shaped like male genitalia includes the lizards mentioned earlier. Pliny the Elder advised that "*its muzzle and feet, taken in white wine, are aphrodisiac, especially with the addition of ragwort and rocket seed*" taken in a small amount. This group also includes the rhino horn. This is essentially just a keratinous powder that contains not a smidgen of anything that could be considered an aphrodisiac. It basically has the same effect as biting your nails, if you discount the placebo effect that can be attributed to rhino horn. Incredibly, because of the severe punishment for poaching these animals in Africa, suppliers of this material have gone so far as to kill rhinos in European zoos.

Obviously, there were – and still are – countless recipes for boosting the libido and we have not even mentioned plant-based aphrodisiacs. A highly prized article of trade in antiquity was terebinth balm, a resin made from the *Pistacia terebinthus*. More affordable plant sources included garlic blended with coriander stirred into wine, and Cyprian reed also mixed into wine. Clematis leaves can also be an aphrodisiac when macerated in wine. Sexual desire can also be achieved with xiphium stem – steeped in wine. Could it have been mainly the wine?

Some part of these plants was often shaped like human genitalia, giving them an aphrodisiac effect; that is why the recipes often called for the tubers and stems, including orchid stems. The name *Orchidaceae* is actually derived from the ancient Greek word for testicle, while Spanish conquistadors coined the name *vanilla* because the pods resemble a vagina. Vanilla is also considered an aphrodisiac. Then there is the arum flower, reminiscent of a penis.

From the late 18th century to the early 20th century, alchemy and quackery offered to cure all sorts of human ailments, including impotence. There was a plethora of various lozenges, potions, tinctures, and syrups. *Dr. Brodum's Botanical Syrup and Restorative Nervous Cordial* promised to prepare men to perform their marital duties. Ebenezer Sibly offered his *Reanimating Solar Tincture* for the same problems, while Samuel Solomon had his *Cordial Balm of Gilead*. There was also *Dr. Senate's Balm of Mecca*, and *R&L Perry's Cordial Balm of Syriacum*. Later analyses showed that most of these contained a mixture of gentian, cardamom, calumbo, and sometimes Spanish fly and all were steeped in strong alcohol, usually brandy. While they did little to help, if you again discount any placebo effect, they did little harm. Some of the syrups may even have tasted quite good.

Chemistry and pharmacy were not alone in the treatment of impotence: physics had its role too, especially electricity. Of all the uses of electricity, probably the least bothersome was Dr. Graham's *Celestial Bed*. Another method that was far more complex required attaching electrodes to the penis and testicles. John Malkovich complained to Bruce Willis of a similar procedure in the movie *RED*, although that had nothing to do with the treatment of impotence and more to do with the Russians wanting to obtain some important information. Even that was nothing compared to the method of inserting electrodes into the urethra and sending in some jolts of electricity, like when your car battery is dead and you use jumper cables to start it up.

Apart from electricity, impotence was also treated via the vacuum method. This involved a vacuum device, or pump, and assumed that the negative pressure would make blood rush to the penis and cause an erection. The first known device was designed in the late 19th century by Eugen Sandow, a German bodybuilder and vaudeville performer. He called it the *peniscope*. In 1913, Dr. Otto Lederer was the first to patent a vacuum device and although there have been tweaks in the materials and design, these types of devices are still manufactured today. Even in the age of Viagra, there are several hundred devices available, ranging in price from $20 to $500. The manufacturers claim a satisfaction rate of 50–80 percent in men who have tried a vacuum pump.

If there is one thing that men can complain about to Mother Nature, it is that they have no penile bone. Most animals do, even bats. Whales have one that can reach up to seven feet in length. Most other primates also have one, including gorillas and chimpanzees. Humans seem to have lost it somewhere in the course of evolution around 50 million years ago, so it seems only natural that they would try to rectify this deficiency if possible.

The first attempts had nothing to do with treating impotence, however. They were needed to address a much more serious problem confronting surgeons during World War II: how to reconstruct the genitals of soldiers injured in combat. The most common form of implantation used rib cartilage, but the implant soon lost its original shape and was slowly absorbed into the patient's body. Artificial materials were needed. The first post-war experiments with acrylic rods were a failure; success would not come until the 1970s. Silicone rods came next, followed by silicone-sponge prostheses, which after implantation absorbed blood and caused the penis to increase in size. The drawback was that the increase was permanent and the penis never returned to a flaccid state. Inflatable prostheses saw the greatest success. These consisted of silicone cylinders implanted into the penis that were controlled through a small pump placed in the patient's scrotum. The whole thing may seem complicated, but the prostheses became pretty popular and 250,000–300,000 of them were implanted in the decade following the discovery.

This was a time when it was assumed that the inability to achieve an erection was more of a psychological than a physiological problem. The problem was in the patient's head, not his genitals. In fact, the official ratio was 80:20, where 80 percent of the problem was considered psychological and only 20 percent was physiological. It was also thought that male impotence was not just a problem for the man, but a couple's problem that could only be treated with the participation of both sex partners. With this insight, clinics cropped up all over the United States, and some elsewhere, to address the issue of erectile dysfunction through couples therapy. It was not the cheapest therapy – the price tag for two weeks in a high-end facility was $2,500 – but it was relatively effective. The best facilities had a 75 percent satisfaction rating.

But let us return to Las Vegas so we can clarify what really happened on that evening in May 1983.

First of all, the man who dropped his pants that night in May 1983 was no exhibitionist. He was also not a Las Vegas freak show, although his behavior makes it seem otherwise. Giles Brindley was in fact a serious scientist, a professor at the University of London, a noted British physiologist, and the author of more than a hundred scientific papers. He was also a musicologist and a composer.

Secondly, the auditorium where his memorable performance took place was not filled with perverts and sexual deviants. It was filled with attendees at the 78th annual meeting of the American Urological Association.

Thirdly, the reason he dropped his trousers was for a demonstration that was part of a presentation on the vasoactive treatment of erectile dysfunction. Tales of elixirs of youth, the Toggenburg goat, and all the amazing things men are willing to endure, are clear evidence of the chimeric efforts in those days to return virility in cases where it was absent. But the watershed moment came in the 1980s. The first to make the discovery was Ronald Virag, a French surgeon and the son of a professional soccer player. During an operation, he accidentally injected papaverine – an antispasmodic drug – into the wrong artery, the artery leading to the penis. To his surprise, it produced an erection in the still-anesthetized patient that lasted a full two hours.

Dr. Giles Brindley had a similar experience, but with a different agent. His colleagues understood that he had achieved significant success in the treatment of impotence with his approach, but they did not have much faith in him. How could they, after all his other "fool-proof" recipes over the years. So the bold scientist did something wild and went down in the history of sexual medicine. The story is best told by eyewitness Dr. Laurence Klotz, who today is a professor of urology at the University of Toronto:

"About 15 minutes before the lecture I took the elevator to go to the lecture hall, and on the next floor a slight, elderly looking and bespectacled man, wearing a blue track suit and carrying a small cigar box, entered the elevator. He appeared quite nervous, and shuffled back and forth. He opened the box in the elevator, which became crowded, and started examining and ruffling through the 35 mm slides of micrographs inside. I was standing next to him, and could vaguely make out the content of the slides, which appeared to be a series of pictures of penile erection. I concluded that this was, indeed, Professor Brindley on his way to the lecture, although his dress seemed inappropriately casual.

"The lecture was given in a large auditorium, with a raised lectern separated by some stairs from the seats. This was an evening programme, between the daytime sessions and an evening reception. It was relatively poorly attended, perhaps 80 people in all. Most attendees came with their partners, clearly on the way to the reception. I was sitting in the third row, and in front of me were about seven middle-aged male urologists and their partners, in 'full evening regalia.'

"Professor Brindley, still in his blue track suit, was introduced as a psychiatrist with broad research interests. He began his lecture without aplomb. He had, he indicated, hypothesized that injection with vasoactive agents into the corporal bodies of the penis might induce an erection. Lacking ready access to an appropriate animal model, and cognisant of the long medical tradition of using oneself as a research subject, he began a series of experiments on self-injection of his penis with various vasoactive agents, including papaverine, phentolamine, and several others. His slide-based talk consisted of a large series of photographs of his penis in various states of tumescence after injection with a variety of doses of phentolamine and papaverine. After viewing about 30 of these slides, there was no doubt in my mind that, at least in Professor Brindley's case, the therapy was effective. Of course, one could not exclude the possibility that erotic stimulation had played a role in acquiring these erections, and Professor Brindley acknowledged this.

"The Professor wanted to make his case in the most convincing style possible. He indicated that, in his view, no normal person would find the experience of giving

a lecture to a large audience to be erotically stimulating or erection-inducing. He had, he said, therefore injected himself with papaverine in his hotel room before coming to give the lecture, and deliberately wore loose clothes (hence the track-suit) to make it possible to exhibit the results. He stepped around the podium, and pulled his loose pants tight up around his genitalia in an attempt to demonstrate his erection.

"At this point, I, and I believe everyone else in the room, was agog. I could scarcely believe what was occurring on stage. But Prof. Brindley was not satisfied. He looked down skeptically at his pants and shook his head with dismay. 'Unfortunately, this doesn't display the results clearly enough.' He then summarily dropped his trousers and shorts, revealing a long, thin, clearly erect penis. There was not a sound in the room. Everyone had stopped breathing.

"But the mere public showing of his erection from the podium was not sufficient. He paused, and seemed to ponder his next move. The sense of drama in the room was palpable. He then said, with gravity, 'I'd like to give some of the audience the opportunity to confirm the degree of tumescence.' With his pants at his knees, he waddled down the stairs, approaching (to their horror) the urologists and their partners in the front row. As he approached them, erection waggling before him, four or five of the women in the front rows threw their arms up in the air, seemingly in unison, and screamed loudly. The scientific merits of the presentation had been overwhelmed, for them, by the novel and unusual mode of demonstrating the results.

"The screams seemed to shock Professor Brindley, who rapidly pulled up his trousers, returned to the podium, and terminated the lecture. The crowd dispersed in a state of flabbergasted disarray. I imagine that the urologists who attended with their partners had a lot of explaining to do. The rest is history."

Some have called this the lecture that changed sexual medicine. Of course, Brindley's manner of demonstration was more than unusual, but more important than that is the fact that for the first time in history, doctors had a new method of treating impotence that was not dependent entirely on the placebo effect. For the first time in history, they could really treat impotence. To paraphrase Neil Armstrong's famous quote in this context, the short journey taken by Professor Brindley from the podium to the astonished audience was one small step for man, one giant leap for the treatment of impotence.

After the lecture in Las Vegas, Professor Brindley continued to research agents capable of inducing an erection. He published a list of 17 such drugs, seven of which did in fact induce an erection.

We can probably guess whose penis was used as the test subject in his studies.

Story 9.5: The big medicine producer and the farmer's beautiful daughter

The penis is truly an exceptional organ. It is unique in that it has no subcutaneous fat. Men have this type of fat throughout their entire bodies except for the penis, and when they put on weight, they put it on everywhere. One may have to get a larger suit, shirt, undergarments and even shoes and gloves. This situation is different with condoms, as the same size, if needed, fits the man his entire life.

The penis is also exceptional in its ability to increase in size and hardness (also known as an erection). The logic behind this activity is quite clear. Only the erect organ is capable of penetrating the vagina and impregnating a woman. But then we have to ask ourselves, why do men not have an erection 24/7? All of the complex psychological and physiological mechanisms necessary to reproduce would then not be necessary at all. On the other hand, our predecessors in evolution would have found it relatively uncomfortable to go around with an erect penis all the time. Moving from place to place and fighting off predators was much easier and safer with a relaxed penis hidden beneath their loincloth, which means there must be some sort of mechanism which makes it possible for the penis to exist in two diametrically different states, both relaxed and erect. This mechanism is the erection.

We have long known about how our individual organs work. The basic physiologies of the heart, lungs, kidneys, and liver were described long before the beginning of the 20th century. To understand the male erection however, we had to wait until the 1980s. The reason was simple. A famous urologist once said, "*...the study of male sexuality was taboo. Cardiovascular studies garnered applause, but the penis? People saw it merely as perversion.*"

Until the end of the 15th century, we believed the teachings of the ancient Greeks that erections were caused by an accumulation of air, a sort of inflation. Air originated in the liver, from where it traveled to the heart and then through the veins to the penis, where it would fill its hollow spaces. The first person to disprove this theory was Leonardo da Vinci. In his efforts to understand the basis of everything, he conducted many autopsies and studied the human body to unimaginable detail, which went against the norms of the time. He was the first person to say that an erect penis is caused by blood filling the organ. Dutch doctor Reinier de Graaf would further study the process of an erection using a male corpse 100 years later. He described in great detail how an erection was achieved by slowly injecting water into the hollow parts of the penis. Then, another 200 years later, it would be discovered that the entire process was controlled by nerve signals being sent from the brain. They learned that an erection is a process where blood is transferred to the hollow parts of the penis and that the signal triggering the change comes from the brain. The question which remained unanswered was, why is there a lot of blood in the penis at one moment and not at another? What causes the blood to remain in the organ and cause the hardness? The answer to this question would finally be found after experiments conducted by Giles Brindley, the hero of the previous story, and his colleague Ronald Virag.

So, how does erection originate? The answer is as follows: Based on a command from the brain, the blood fills the sponge structure surrounding the veins in the penis that are responsible for blood flowing out of the penis. The spongy parts expand, closing the veins to prevent the outflow of blood. The blood remains in the penis, ensuring it will be hard. After an orgasm or an unexpected interruption, another command comes from the brain, the blood leaves the spongy organs and they decrease in size. The arteries outflowing the blood relax and the rest of the blood leaves the penis. If everything functions as it should, a man will always get an erection when he wants it, when his brain sends a signal to start the action. If things are not functioning properly, however, problems with an erection occur. Until recently, this disorder was called impotency, and this was the word we used in our stories until now. But from now on, we will use the new, official name for this disorder: erectile dysfunction.

The term erectile dysfunction appeared approximately towards the end of the 1970s. However, the new designation was not made official by the National Institutes of Health in the United States until December 1992. At first glance, it appears to be merely an exchange of one name for another, but that is not entirely the case. Impotency does not only mean the incapability to have an erection; it did and still does include infertility and a lack of sexual desire, called libido. The name itself also has quite a pejorative meaning. The impact of this confusion, however, is more important. While impotence was a male problem as such, erectile dysfunction narrowed the problem down to one organ, the penis; something like appendicitis (although it is true that the appendix does not have nearly as many names as the penis). Secondly, the name change caused a shift away from the psychological-emotional riddle and allowed doctors to step in and do something about it. This approach is known as medicalization, although this is not a very popular word. It implies that what used to be simply a sort of problem suddenly becomes a medical diagnosis. Doctors then immediately take the diagnosis into their hands, and what was once a healthy person, albeit with a problem, now becomes a patient. This is, in fact, a fairly accurate description. What is surprising about erectile dysfunction is that medicalization did not happen for such a long time. The female body was medicalized much earlier. Menstruation, pregnancy, childbirth, and menopause are all examples of problems that are commonly dealt with by doctors and women are therefore more comfortable communicating their problems to their doctors than men. This did not happen for men until much later. Take urology textbooks, for example. In the third edition of the *Campbell Textbook of Urology*, published in 1970, only a single page dealt with impotency. The fifth edition of the same book, published in 1986, contained 30 pages about impotency, and by the sixth edition, published in 1992, there were 50 pages. In the seventh edition, from 1998, there were 180 pages focused on impotency, but now under the name erectile dysfunction. It was around this time that pharmaceutical giant Pfizer introduced their revolutionary new drug – Viagra – which was approved on March 27, 1998.

Many critics accused Pfizer of inventing the term erectile dysfunction as a marketing strategy, while others admired them for the same thing. But while male impotency was considered a problem – and doctors did not usually deal with problems – erectile dysfunction was a real medical diagnosis. And a medical diagnosis deserves a pharmaceutical solution, Viagra.

Yet this was not completely the case. Both labels, erectile dysfunction as well as Viagra, are almost the same age, but the first one is actually just a little older. The drug manufacturer simply got lucky in its timing, with the diagnosis only being a few years old. The fact of the matter is, Pfizer accepted, occupied and literally took ownership of this diagnosis. Pfizer was lucky again just six months before the drug came to market, when the Food and Drug Administration in the USA overturned its long-time ban on promoting and advertising drugs directly to customers and patients. This now allowed the manufacturer to bypass doctors completely and launch an advertising campaign (many more would follow) targeting end-users. The luck did not end there. Just two months before the drug was put on the market, the sex scandal of President Bill Clinton and intern Monica Lewinsky rocked the country. Prude America now felt free to talk about sex, sex came to prime-time television and it was splashed across magazine covers and dominated newspaper headlines. People discussed it publicly without blushing. These discussions were not only abstract, but also based on completely specific images (Miss Lewinsky kneeling before the president in the Oval Office, he with his legs open…). All in all, Pfizer came to the table with a drug for a completely fresh, but already existing medical problem, erectile dysfunction. It was able to advertise it directly to potential customers, and the idea of sex was now being talked about openly. Now that is what you call a streak of great luck! But only now are we arriving at the biggest serving of luck.

The story of Viagra's origin is rather extensive and there are many different versions. According to the most popular and most widespread version, the active substance sildenafil was originally prepared and clinically tested by Pfizer as a treatment for heart disease. The results were inconclusive, so they decided to end the experiments and forget about the drug. However, after checking the tablets, the research organizers realized that many of the patients had not returned the unused drugs. When asked why, the patients all replied that the drug gave them an erection. This realization changed the purpose of the clinical experiments and, given that the patients' claims had been confirmed statistically, a new drug had been discovered.

Well, this story is almost true. What is definitely not true is that the patients were able to keep the drugs. In clinical experiments of this sort, that is simply not possible. The drugs are strictly monitored by doctors and it is out of the question that some of the drugs would turn up missing after the experiments.

The compound that would later become the active substance of Viagra was synthesized in the British laboratories of Pfizer at the end of the 1980s. It was given the code name UK-92,480 and was really intended for the treatment of high blood pressure and ischemic heart disease. The first phase of the clinical experiments was undertaken in a hospital in Morriston, Wales. It was a short experiment, with the drug being given to healthy volunteers. The main point of the investigation was not about the effectiveness of the substance, but rather to determine its safety, side effects, and behavior in the human body. In a report from one of the experiments, one patient reported gastrointestinal, back and leg pain among the side effects. It was also noted that "*There were also reports of erections.*" Nothing else. Moreover, at the dose used in the study, reports of erections were rare and, if it did occur, it was days after the drug was administered. Because of this, Pfizer decided to continue testing UK-92,480 as a medicine for heart problems. Who would be interested in a drug that you needed to take on Wednesday so that you could get an erection on Saturday?

Over time, they found that the substance would not be useful in treating heart disease, but there was an increase in the number of reports of erections as a side effect of the drug. The manufacturer then decided to conduct a small pilot study on patients with erectile dysfunction. The plan of the study was quite simple. One group of patients received the actual drug while a second group received an ineffective placebo. Both groups of men were then shown an erotic video, during which a special machine measured the circumference and hardness of the penis. The results were more than satisfactory. The drug was substantially more effective than the placebo.

Still, this was not enough to decide whether Pfizer should go in this new, unexplored direction. The company still faced the main part of the experiments involving the long-term administration of the drug to thousands of patients throughout the world. The costs ran into the hundreds of millions of dollars, the odds of success were around 1 in 5, and the testing was done in a field no one had any experience with. But Pfizer decided to continue with the clinical studies of the substance known as UK-92,480 as a treatment for erectile dysfunction. The rest of the story is well known.

Viagra was really effective for the treatment of erectile dysfunction. Its effect can be explained in that it relaxes the smooth musculature from which the spongy organs around the veins that outflow blood from the penis are formed. These muscles must be in a relaxed state in order to allow blood in and thus increase their volume. If it does not happen, the miracle of an erection cannot take place. Thanks to Viagra, these muscles relax, letting in blood, increasing the size, preventing blood from leaving the penis, and letting the penis harden, to the satisfaction of its owner.

Viagra's entrance onto the pharmaceutical market was truly triumphant. The FDA approved the drug on March 27, 1998 and it was on the shelves of pharmacies across America the following month. In just the first four weeks alone, doctors prescribed it 300,000 times, and by the end of the first quarter it had been prescribed 2.7 million times. Turnover for the first quarter was $400 million, and the end of the first year saw it hit the magic *billion* number. Interest in Viagra did not diminish at all in the United States, as evidenced by the fact that between 2003 and 2016, sales averaged $1.7 billion a year.

There was no lag in other countries either, not even where the advertisement campaigns were significantly more conservative than in the U.S. During the first ten years after Viagra's release, British men had consumed 37 million tablets of the new drug.

It would not be correct to correlate Viagra's success only to luck. Many world-class experts from over 100 different scientific and professional disciplines were involved in the development and presentation of Viagra, but like any good business, luck played a role as well. As Julius H. Comroe, Jr., President of the American Physiological Society, once said, "*Serendipity is looking in a haystack for a needle and discovering a farmer's daughter.*"

Concluding remarks

The name Viagra is a portmanteau combining vigor and Niagara. It suggests vitality and strength, which can also be considered attributes of the famous waterfalls on the border of the U.S. and Canada. Viagra certainly lives up to its name. The public was convinced of its efficacy as soon as it came on the market, turning it into the fastest selling drug in history, to the delight of millions of men around the world. Arnold Melman, Chief of Urology at *Montefiore Medical Center* in the Bronx, says "*I see a lot of municipal workers in my practice. These are tough, physical guys – bus drivers, subway motormen, laborers, etcetera. I've gotten used to seeing them cry in my office. They cry twice, in fact: first, when they tell me they can't get an erection; and second, after we treat them, when they tell me they can.*"

It was not just patients with erectile dysfunction that were thrilled with the blue pill. Pfizer shareholders were no less ecstatic. Even before the drug was put on the market, the expectations alone caused the stock price to double. The growth continued, and between April 1998 and April 1999, it increased another 50 percent.

Apart from having a huge financial and medical impact, Viagra also became a cultural and cult phenomenon of the late 20th and early 21st centuries. The introduction of Viagra opened the floodgates for medical, psychological, and sociological studies. Some say that Viagra kicked off the second sexual revolution. Whether or not that is true, one thing is certain: Viagra caused a substantial shift in how sex is perceived, particularly in the lives of men. Now they can do it whenever they want. This has caused some women authors to criticize it as another step toward consolidating male dominance, while men praise it as breaking free of feminist oppression.

The popularity of Viagra is also evident in the fact that it has become the most counterfeited drug ever. It is estimated that up to three-quarters of all the Viagra sold on the internet is fake. This poses a serious health risk. The fake Viagra is so profitable that even drug dealers have started to switch from producing and distributing cocaine to these blue pills.

Not only does Viagra help men, it is also helping to save endangered species. At one zoo in China, they give Viagra to pandas to help them reproduce. Viagra is also saving other species, like seals and tigers. Seal genitals and tiger penises were once commonly used to increase sex drive. After the discovery of Viagra, the price of the first dropped on the Chinese market by 20 percent from the original $100. Chinese men simply prefer Western advances to their own traditional preparations and that is saving seals and tigers.

That is no joke – unlike the 944 Viagra jokes told by Jay Leno on the *Tonight Show*, according to one count. Jay Leno was not the only one telling Viagra jokes. No drug other than the one for erectile dysfunction has ever triggered such an avalanche of humor. Take *The Big Viagra Jokebook* – 186 pages containing nearly 350 Viagra jokes. Here is one for your enjoyment:

Do you know what real hard liquor is? Whiskey mixed with Viagra.

10

Vaccines

"Before the Spaniards appeared to us, first an epidemic broke out, a sickness of pustules. It began in September. Large bumps spread on people; some were entirely covered. They spread everywhere, on the face, the head, the chest, etc. They could no longer walk about, but lay in their dwellings and sleeping places, no longer able to move or stir. They were unable to change position, to stretch out on their sides or face down, or raise their heads. And when they made a motion, they called out loudly. The pustules that covered people caused great desolation; very many people died of them, and many just starved to death; starvation reigned, and no one took care of others any longer. On some people, the pustules appeared only far apart, and they did not suffer greatly, nor did many of them die of it. But many people's faces were spoiled by it, their faces and noses were made rough. Some lost an eye or were blinded."

That is an excerpt from *General History of the Things of New Spain (Historia General De Las Cosas De La Nueva España)*, written by Franciscan priest Bernardino de Sahagún, in which he describes an epidemic of smallpox during the time of the conquest of what is today Mexico. It broke out in the 16th century, but it was not the first or the last of all the various epidemics and historical literature contains countless similar descriptions. The numerous epidemics of smallpox, cholera, typhoid, tetanus, diphtheria, poliomyelitis, yellow fever, and many other diseases have been decimating civilizations since time began.

One of these diseases has been completely eradicated, and the others are more or less under control, at least in terms of Western civilization. This can be attributed to improved hygiene, a higher standard of living, better healthcare, and many other factors. One of the greatest discoveries in the history of medicine certainly played a significant role – the vaccine.

V. Marko, *From Aspirin to Viagra*, Springer Praxis Books,
https://doi.org/10.1007/978-3-030-44286-6_10

Vaccination, immunization, inoculation. These words are often used interchangeably, so let us make some sense out of them.

We will start with the term *immunity*. The simple definition is the ability of an organism to produce antibodies that help it to resist microorganisms, toxins and viruses. It can be innate or adaptive.

Innate immunity is genetic and we get it from our parents. Because we are the descendants of generations of ancestors who mostly were able to survive all sorts of diseases and did not die of them – at least not before producing their offspring who then became the next generation of our ancestors – some of the resistance they developed was passed down to us. That is one explanation as to why epidemics are no longer as frequent or widespread and devastating as they were in the past. It is also why they were so fatal for the original inhabitants of the Americas. We will talk about that in the third story.

Adaptive immunity is created throughout life, either naturally or artificially. Adaptive immunity is created naturally through contact with infection, typically with an infected person. In the past, this was the only way immunity could be created. If a person survived a smallpox epidemic, they would be immune to any future occurrences. Naturally-created adaptive immunity is still used as a method. The author remembers how his mother would make him and his brothers (there were four boys) get into bed with whichever one of them had the measles or chicken pox during childhood. They would all get sick and their mother dealt with all of them at the same time; she did not have to worry that one of them would get the disease in the future. They all developed an immunity to these diseases.

The procedure by which immunity is created is called immunization. The author and his three brothers were immunized with the natural method. Today, the most common form of immunity creation is through targeted immunization, the deliberate introduction of an immunogenic substance into a healthy person. Targeted immunization is known as vaccination and the immunogenic substance is known as a vaccine. The second story describes how the vaccine got its name.

The first known method of targeted immunization was variolation, which is mentioned in the first story. It was named after the virus that causes smallpox: variola. This was a simple procedure used to prevent smallpox in the 18th century, and it consisted of taking material from patients and using it to immunize healthy individuals. It made the person sick, but the case would be mild and the entire point was that the person would be immune to smallpox in the future. Even though this method for preventing smallpox was successful, it was still a risky procedure. Variolation was later replaced with safer methods of vaccination.

Finally, there is inoculation, the oldest method of vaccination. Originally this consisted of creating a small wound, usually in the arm, into which a modest amount of vaccine was applied. The scar was frequently in the shape of an eyelet and "eye" in Latin is *oculus*.

There are many more forms of vaccination available today. Vaccine can be introduced into the body, for instance, by injection, by swallowing, or by inhaling.

Story 10.1: A beautiful aristocrat and the Ottoman method

"Apropos of distempers, I am going to tell you a thing that will make you wish yourself here. The small-pox, so fatal, and so general amongst us, is here entirely harmless by the invention of ingrafting, which is the term they give it. There is a set of old women, who make it their business to perform the operation, every autumn, in the month of September... People send to one another to know if any of their family has a mind to have the small-pox: they make parties for this purpose, and when they are met (commonly fifteen or sixteen together) the old woman comes with a nutshell full of matter of the best sort of small-pox, and asks what vein you please to have opened. She immediately rips open that you offer to her, with a large needle, (which gives you no more pain than a common scratch) and puts into the vein as much matter as can ly upon the head of her needle, and after that, binds up the little wound with a hollow bit of shell; and in this manner opens four or five veins. The children or young patients play together all the rest of the day, and are in perfect health to the eighth. Then the fever begins to seize them, and they keep their beds two days, very seldom three... and in eight days' time they are as well as before their illness. Every year thousands undergo this operation... there is no example of any one that has died in it..."

These are excerpts from a letter sent by Lady Mary Montagu from Constantinople (now Istanbul) to her friend Sarah Chiswell in London. The letter is dated April 1, 1717. At the time, Constantinople was the capital of the Ottoman Empire and Lady Mary was there as the wife of the British ambassador, Edward Wortley Montagu. Although the possibilities of combating smallpox had already been mentioned somewhat earlier in academic literature, Lady Mary's letter of April 1, 1717 is considered the milestone in the battle against this devastating disease.

Lady Mary Wortley Montagu was an extraordinary woman and she most definitely deserves her role as hero of this story. But first we should look at the disease itself. It was once one of the most feared diseases, one that we are now, thanks in part to Lady Mary, only familiar with through stories in literature.

Historically, smallpox was one of the most frequently occurring and deadliest of the infectious diseases. It was a part of human history for centuries, if not millennia. It was caused by a virus called variola, which is exceptional in that it only infects humans. That means it could not be transmitted by other animals or insects as many other infectious diseases can be, it could only be transmitted from human to human. When the first symptoms appeared in a patient, millions of viruses began forming in their mouth and throat. These found their way into the saliva and were dispersed from the patient's mouth when they spoke or coughed. Anyone in the vicinity could breathe them in and become another link in the infectious chain.

The infected person remained symptom-free for the first 7–10 days and usually felt quite healthy, but during that time the virus multiplied and took root in the organism. The first symptoms were chills and a high temperature, followed by a severe headache and backache. Some people fell into delirium or had spasms. The fever and other symptoms subsided after a few days, but soon small red sores would begin to appear in the mouth and on the face, spreading all over the body. They were most concentrated on the face and limbs. The patient would begin having trouble swallowing and the sores in the mouth and throat would grow larger and begin to fill with white fluid, eventually becoming pustules. The pustules would become deeply embedded in the skin, causing indescribable pain. There were thousands of them, sometimes covering practically the entire face. They continued to grow for around two weeks before they started to scab over. Often, only seven of ten patients survived this phase. Those who were lucky saw the scabs start to flake off, their other symptoms would gradually recede and they stopped being infectious. Unfortunately, the scabs left permanent deep scars on the face and body, and some of the survivors were blinded. In fact, smallpox was the chief cause of blindness in Europe in the 17th and 18th centuries. Regardless of any lasting effects, those who survived were immune to any subsequent outbreaks of the disease.

Since the virus was only transmitted from human to human, a certain population density was required for infectious outbreaks, so it comes as no surprise that smallpox first appeared in the ancient centers of civilization in Egypt and southeast Asia. The first notable case was the Egyptian pharaoh Ramesses V, whose reign lasted only four years somewhere around 1100 BCE. Not much is known about his life, but his mummified remains were preserved. There were obvious smallpox scars on his face and arms.

Even older descriptions of smallpox were preserved in ancient India. There is even a Hindu goddess of smallpox named Shitala Mata, indicating just how seriously the disease was viewed by the inhabitants.

In Europe, smallpox is first mentioned as the cause of the Athenian plague in 430 BCE. The epidemic lasted two or three years and caused the deaths of 25 percent of Athens' soldiers and a similar number of the general population. The Antonine Plague that afflicted ancient Rome from AD 165–180, causing around five million deaths, is also thought to have been caused by smallpox. Waves of epidemics wiped out city and country dwellers in the Middle Ages too. Smallpox may have been one of the reasons why the Huns retreated from Gaul in the fifth century. Thanks to the Huns, the Christian world also has a patron saint of smallpox. It is Saint Nicaise, who survived smallpox when he was the bishop of Rheims. His execution by the Huns later led to his canonization.

As the population grew, so did the cases of smallpox. Armies and merchants spread the disease throughout the civilized world. In the tenth century, smallpox was known in China, Japan, southeast Asia, and the Mediterranean coast. When it first appeared in Iceland in 1241, it killed 20,000 of the total population of 70,000. In the 18th century, smallpox killed 400,000 Europeans every year, including monarchs such as France's King Louis XV, Mary II of England, William II, Prince of Orange, and Russian Czar Peter II.

There was no cure for smallpox. Nothing worked; not herbs, not bloodletting, not sweating in an attempt to expel the toxins from the body. But at the beginning of our story we mentioned that some parts of the world had long known of a way to avoid this fatal disease. Lady Montagu wrote about it in her letter and it was later given the name variolation. It was first described in China in the 15th century and somewhat later in Sudan. In all likelihood, it found its way from one of those countries to what is today Turkey, where the wife of the British ambassador became familiar with it in the early 18th century. From there, it came to Europe courtesy of this beautiful Englishwoman.

Lady Mary Wortley Montagu was described as a beautiful woman by her contemporaries, despite the fact that she contracted smallpox at the age of 26, which ravaged her face and caused her eyelashes to fall out. She was definitely one of the most charming women of her day, intelligent, educated and independent. She hated boredom and foolishness and did everything she could to avoid them. She was born in May 1689 into an aristocratic family, the daughter of the 1st Duke of Kingston-Upon-Hull in Yorkshire. Her mother passed away when Mary was only four years old, and for a time her grandmother cared for her and her three younger siblings. After her grandmother's death, everything fell on the shoulders of nine-year-old Mary.

She enjoyed learning from a very young age, and her father's large library was her favorite place – the place where she secretly obtained an education that was reserved strictly for men at the time. As a result, she knew Latin by the time she was eight and she had written two collections of poetry and an epistolary novel by the time she was 14.

Mary had two serious suitors at the age of 21. One was chosen by her father and the other she chose for herself: Edward Wortley Montagu. She resolved the problem by eloping with Edward. That left her father with no option but to accept her choice.

Her husband was soon given a post at court in London, and Mary and their young son moved to the capital. Her wit and beauty made her a hit in the court of King George I, but her life was not all happiness and joy; her younger brother died of smallpox at the age of 20. Lady Mary fell ill with the disease two years later, and although she was luckier than her brother, the disease left deep marks on her beautiful face.

Her husband's appointment as ambassador to the Ottoman Empire changed Mary's life significantly. He held the post only briefly, less than two years, but even this short amount of time considerably influenced her life. She gave birth to a daughter in Istanbul, but mainly she had access to the society of Ottoman women, which was off limits to men and about which, until then, only limited and distorted information had been available. Mary Montagu was the first member of Western civilization to be able to provide accurate information about the life of women in the Orient. In countless letters, she shared her unusual insights and experiences with her friends back home in England.

This brings us back to the main subject of this story – variolation, the name later given to the method she described in her letters. It consisted of introducing a small amount of the smallpox virus into the human body by inoculation into a small wound created on the skin. The inoculation material was obtained from the pustules of a person infected with smallpox. The amount had to be just small enough not to induce the most severe symptoms of the disease, but sufficient for the body to develop immunity to the disease and for that immunity to protect the person when subsequently exposed to the virus during another of the numerous smallpox epidemics.

Lady Mary Montagu already knew the effects of smallpox. When she was convinced of the effectiveness of variolation, she did not hesitate to have her four-year-old son inoculated. She actively promoted the Ottoman method after she returned to England, but she was met with great resistance from the medical community. Physicians rejected a procedure that was based solely on the knowledge of Oriental folk medicine, but Mary was determined and persistent and ultimately wore down the skepticism. The turning point was when she had her young daughter inoculated under the supervision of the physicians to the court. The young Edward Wortley Montagu, Jr. and little Mary Wortley Montagu became the first known Europeans to be inoculated against smallpox.

The first people to undergo a sort of clinical trial of the new procedure were six prisoners at Newgate Prison in London. These volunteers were willing to be inoculated with smallpox and then be exposed to the actual infectious disease. There was some risk involved, to be sure, but they were promised their freedom for taking part in the "clinical trial." If they survived, that is. The trial took place in 1721. The prisoners survived, the first inoculation was a success, and the variolation method could begin its victorious campaign around England.

Variolation not only spread across England, but the country ultimately became something of an international center of inoculation. The simple method of smallpox prevention spread from there to the entire world. Frequently, it was a smallpox epidemic in any given country that would force the residents to try the prevention method as a way of protecting themselves from the disease. For instance, variolation arrived in France in the second half of the 18th century after the 1752 smallpox epidemic in Paris nearly killed the heir to the throne. A smallpox outbreak in Nagasaki around the same time led to the method being welcomed to Japan. The procedure became widespread in America during the American Revolutionary War, after George Washington ordered the entire Continental Army to be inoculated.

It is worth noting the manner in which variolation was introduced into Russia. In 1762, Catherine the Great invited prominent London physician, politician, and banker (a rather unusual combination of professions for one person) Thomas Dimsdale to inoculate her son, the heir to the throne who would later become Czar Paul I. Preparations took quite a long time, but in 1768 Dimsdale inoculated not only the czarina and her son, but also another 140 members of the court. Since the outcome was uncertain, a relay of horses was at the ready to whisk him away as fast as possible from St. Petersburg. The inoculations were successful and the reward was truly princely: £10,000 in cash, an annuity of £500, and the title of baron for both Thomas Dimsdale and his son. For comparison, the annual salary of an unqualified worker at the time was less than £25, while one night in a posh house of ill repute, including dinner, bath, and courtesans, would have cost the baron Dimsdale upwards of £6, were he of the mind.

Variolation was the first effective method of protection from the deadly virus. It was simple and the benefits were felt during every smallpox epidemic – which occurred quite frequently in the 18th century. The difference in morbidity in those who were inoculated compared with the rest of the population was seen immediately. When the first variolation took place in Boston in 1721, only six of the 300 inoculated patients died of smallpox in the epidemic. As a comparison, from 6,000 people that had contracted the disease in that time, 1,000 died out.

Despite the obvious advantages of variolation, it was very slow to spread, primarily for religious reasons. Smallpox was still frequently considered to be a punishment from God and variolation was thought to be blasphemous because there was no mention of it in the Bible, and because it went against God's right to decide about the life and death of ordinary mortals. Alongside this, empirically-minded critics of variolation could not understand the use of a virus to prevent a virus.

We have seen that variolation was the first method of protection from smallpox. As with most first methods, it was not without certain problems. A healthy person was inoculated with a deadly virus in order to develop immunity to subsequent infections caused by the same virus. The description alone suggests that the procedure carried a certain risk. The vast majority of infected individuals survived immunization and most of them did develop immunity but there were a handful of unlucky individuals, including one of the sons of King George III, for whom the procedure failed and they died of the disease to which they had been introduced. Variolation more or less continued until the late 18th century, when it was gradually replaced by a safer method of immunization – vaccination.

And what of Lady Mary Wortley Montagu and her turbulent life following her return from Constantinople? She became a friend and the muse of Alexander Pope, one of the greatest English poets. She rejected his courtship and the great poet responded by attacking her in his literary works. At the age of 47, while still married, she fell in love with 24-year-old Count Francesco Algarotti, an Italian polymath, philosopher, poet, essayist, art collector, and one of the most dashing gentlemen of his time. She left England to stay with him in Venice. Count Algarotti was also bisexual, which meant Mary also had to compete with the English politician John Hervey. Her love for the count and their travels together around Europe lasted about five years. When their relationship ended, she left sunny Italy for sunny Provencal, ultimately returning to England shortly before her death.

Lady Mary definitely did not suffer from a boring life. Someone once wrote about her: "*She had lived her life in a richly personal way... never bothering to be fussy or correct, and never influenced by public judgment... she knew what she wanted and went and took it without apologizing.*" Life with her was probably not easy, but if she had not been the way she was, history would have been deprived of a truly interesting person and we would likely have had to wait a little longer for the first inoculation against smallpox.

Story 10.2: A wise farmer, a famous doctor and how vaccination got its name

One observation in the history of smallpox proved to be extremely important: it was not just those who had survived a bout of smallpox that were immune to the disease. So were some people who had never had it before, specifically, farmers and farm workers who came into contact with cows. Milk maids had the greatest resistance to smallpox, so much so that during epidemics they could take care of the sick without being at risk of contracting the deadly disease. The only thing they had to do was first contract cowpox which, as the name implies, is a disease that afflicts mostly cattle. However, it can pass from animal to human, especially by touching the lesions that form on the udders of sick cows, and that is the reason why milk maids were primarily infected with the disease. The symptoms are similar to smallpox, but cowpox is much milder. The most important thing for this story, and for humankind, is the fact that contracting this "mild" pox virus protected a person from the related but deadly one.

This observation was not completely new in 18th century England, but the first to take the bold step and actively use it was a wise English farmer.

Benjamin Jesty was a prosperous farmer living in Yetminster in the south of England. He owned a large estate called Upbury Farm and was a pillar of the community. At the time of this story, in 1774, he was 37 years old and had been happily married for four years. He and his wife Elizabeth had two young boys and a newborn daughter. Benjamin ran his family and his farm with a firm, decisive hand, but he was also an astute observer and was able to connect the dots where things appeared to be unrelated. He noticed that when one of the many smallpox epidemics broke out in the area, neither he nor his milk maids, Anne and Mary, contracted the disease. He remembered that he had come down with cowpox as a young boy and that both of the young women had also had the disease. It occurred to him that his wife and two sons might be protected from smallpox if they contracted cowpox, but in their case it would be a deliberate rather than an accidental infection.

Jesty decided to take action when another smallpox epidemic hit the Yetminster area. He took all three members of his family to the pasture of a neighboring farm, which was less than two and a half miles away so they walked. Jesty found a cow with obvious symptoms of cowpox and with a large darning needle, he scratched the arms of his wife and sons and transferred infected material from the pustules on the cow's udder to the scratches he had made on their arms. When the operation was finished, they all returned home to Upbury.

The boys had only a mild local reaction and were soon in good health again but Elizabeth had a worse reaction; in fact, it was serious and for a while Jesty was worried she might die. She ultimately recovered and the farmer's experiment could be deemed a success.

The operation was well-planned and Benjamin Jesty had to have known about the risks involved, but what he could not have known was how the neighbors would respond. The deliberate transfer of a disease from animal to human was considered disgusting and they called him a monster. He was reviled and was insulted and cursed wherever he went.

But time healed not only the scratches on the arms of his wife and sons, but gradually even the hostility of Jesty's neighbors, particularly when it was shown over time that neither his wife nor his sons ever contracted smallpox.

The remainder of Benjamin Jesty's life was unremarkable. In 1797, he moved his family to Worth Matravers on the Dorset coast. His good life had caused him to gain a lot of weight, and he died in 1816 at the age of 79. His wife had the following inscribed on his headstone: "*To the Memory of Benj.in Jesty (of Downshay) who departed this Life, April 16th 1816 aged 79 Years… an upright honest Man: particularly noted for having been the first Person (known) that Introduced the Cow Pox by Inoculation, and who from his great strength of mind made the Experiment from the (Cow) on his Wife and two Sons in the year 1774.*"

His courageous act was nearly forgotten.

Fortunately, that did not happen. On May 17, 1749, another Englishman was born in Berkeley, just north of Jesty's birthplace. After his death, he was called "*the father of immunology*" and "*a man whose work saved more lives than the work of any other human*." A man who was named one of the greatest Britons in history. There are monuments to him in Gloucester Cathedral and Kensington Gardens in London. Hospital wards and streets are named after him, as are towns in the United States. A lunar crater was named in his honor. His name was Edward Anthony Jenner.

Edward was born as the eighth of nine children to the Reverend Stephen Jenner, the vicar of Berkeley. He was variolated at the age of eight, and at the age of 14 he began to study medicine. He gained practical knowledge as a surgeon's apprentice and theoretical knowledge at St George's Hospital in London. In 1773, at the age of 24, he became a family doctor in his hometown of Berkeley.

But Jenner was no ordinary country doctor. He was a scientist on the level of the great names of the 18th century, and he was a man of many hobbies. He was interested in geology, conducted experiments with human blood, and researched the disease known today as angina pectoris. As a biologist, he studied the cuckoo and was the first to describe the behavior of newly-hatched fledgling chicks in the nests of its hosts. He also studied bird migration and hedgehog hibernation.

Medicine and natural sciences were not his only interests. Jenner was also one of the first to construct and test a hydrogen balloon. Although this experiment had little impact on his professional career, it had great significance in his private life. He conducted the experiment a year after the Montgolfier brothers, Joseph-Michel and Jacques-Étienne, first demonstrated their hot air balloon. Jenner's balloon was launched from Berkeley Castle on September 2, 1784 at 2:00 pm and flew ten miles northeastwards to land on the property of one Anthony Kingscote. The land-owner had several daughters and one of them, Catherine, became Mrs. Jenner in March 1788.

That milk maids were immune to smallpox was a fact known not only in Benjamin Jesty's county of Dorset. It was also known in Gloucestershire where Edward Jenner lived and they also knew that this immunity required a prior cow-pox infection. The young Edward thought that this protective effect could also be achieved by a means other than a transfer from cows to humans. Perhaps it could also be achieved by transfer from a human with cowpox to a healthy human. It took some time before the first person with cowpox – a milk maid – was found in the area, but it finally happened in May 1796 and Jenner could start to verify his assumption. The whole experiment is described in literature, with all the details.

The young milk maid was Sarah Nelms and she had obvious cowpox lesions on her arms and hands. The drawing of the milk maid´s right hand became a part of the history of medicine and can be found in the National Library of Medicine in England. The cow from which Sarah became infected was named Blossom and its hide is also available for viewing as it hangs on the wall of the medical library at St George's Hospital in London.

That famous day, written about in the syllabus of many medical schools, was May 14, 1796. That Saturday, Jenner inoculated eight-year-old James Phipps, the healthy son of the doctor's gardener, with some pus from the blisters on Sarah's hand. A few days later, the boy began to exhibit mild symptoms of the disease, such as axillary pain, fever, headache, and loss of appetite. But roughly ten days later, he was healthy again. Edward Anthony Jenner then did something that Dorset farmer Benjamin Jesty did not have the facilities to do. Seven weeks later, on Friday, July 1, 1796, he inoculated James with a fresh smallpox infection. The boy did not show any signs of the disease.

The following year, Jenner sent a brief paper with his results to the Royal Society for Improving Natural Knowledge, Great Britain's top scientific institution. They sent his paper back, noting that if the author wished to maintain his scientific reputation, he should refrain in the future from ideas such as the use of cowpox to treat smallpox. Jenner did not give up. He inoculated another five children with cowpox and published the results in 1798 in a booklet he named *An Inquiry into the Causes and Effects of Variolae Vaccinae, a disease discovered in some of the western counties of England, particularly Gloucestershire, and Known by the Name of Cow Pox*. The booklet was met with a mixed reaction in the medical community, but the title of the booklet carries the idea for the name that was later given to the procedure described by Edward Jenner. The word we know today – vaccination – was derived from the Latin words *vacca* meaning cow and *vaccinia* meaning cowpox. Louis Pasteur, one of the biggest names in biology, was the first to use it nearly 100 years later.

After his failed attempt to convince established medical circles of the importance of vaccination, Jenner decided on a different approach. He gave vaccination material to everyone who was interested. Slowly, more and more physicians were convinced that vaccinations were effective and Jenner's method was soon accepted throughout England, and from there it went out into the world.

In 1802, the Duke of York gave the order to vaccinate the British troops, and Napoleon soon followed with his own order. It is said that when Jenner personally requested that Napoleon release some English prisoners of war, Napoleon granted the request with the words: "*Je ne puis rien refuser a Jenner!*" (I cannot refuse anything to Jenner!). Maria Feodorovna, widow of Russian Czar Alexander III, gave Jenner a diamond ring in gratitude for his discovery. She also ordered that the first child in Russia to be vaccinated be given the name Vaccinov. In the late 18th and early 19th centuries, the vaccine arrived in America where Thomas Jefferson, the third president of the U.S., was a big supporter.

Ironically, it was armies and wars that helped spread vaccinations. In times of war, a large number of men from various parts of the country, some of which were sparsely settled, would gather close together all at once in a relatively small area. Many of them had never encountered transmittable diseases and so they were highly susceptible to various infections. Smallpox was one of the most widespread infectious diseases. During the American Civil War, before vaccination was introduced, almost 20,000 men contracted smallpox – and a third of them died. Over time, all new recruits were vaccinated. When the war ended, civilians began to learn about the benefits of vaccination.

The Franco-Prussian War (1870–1871) played the same role in spreading vaccinations in Europe as the American Civil War had across the American continent, and for the exact same reason: large numbers of men in a small space. The great advantage of vaccination became very clear in this particular war. If we look at it from the perspective of how vaccination advanced, then this war could actually be considered something of a huge clinical study. As the name implies, the war was between France and Prussia. At the time, vaccination for the Prussian troops was mandatory. Of the 800,000 men, 8,643 contracted smallpox and only 459 of them died. There were a million soldiers in the French army, where vaccination was not mandatory, and of the 125,000 men who contracted smallpox, over 23,000 died. That is 8,643 vaccinated versus 125,000 unvaccinated who contracted smallpox and 459 vaccinated versus 23,000 unvaccinated who died of the disease. Few clinical studies in history have achieved such significant variances.

Additionally, it was the French soldiers who started the great smallpox epidemic in Europe. France lost the war and many French soldiers became prisoners of war in Germany. Unlike the army, vaccination was not mandatory for German civilians and in the years after the war, large numbers of civilians became infected with smallpox. Over 160,000 died. The infection spread to other European countries and over the course of five years, almost 500,000 people died.

Edward Jenner did not live to see that. As his contribution was gradually recognized, so his prominence grew. The British Parliament awarded him £10,000 in 1802 and five years later he was awarded another £20,000. Unfortunately, he also found himself the object of ridicule and the butt of jokes. Most often, it was caricatures of cow vaccine patients with cow features.

But Jenner was unchanged by either ridicule or honor. He gradually withdrew from public life and went back to his country doctor practice. He built a modest gazebo near Chantry House, where he lived, and there he vaccinated the poor for free. In a short span of time, his oldest son, his sister, and his wife all died of tuberculosis. Jenner passed away at home on January 26, 1823, overwhelmed by his personal losses.

Benjamin Jesty, the wise and courageous farmer, deserves a few last words. After Edward Jenner's contribution to the battle against smallpox was recognized, and shortly after he received the first award of £10,000, the Reverend Andrew Bell, a minister from a town near Jesty's home, wrote a letter stating that Jesty had been the first vaccinator. He sent the letter to a member of parliament and to the director of the newly created Original Vaccine Pock Institute. They summoned Jesty before officers of the institute where he was to prove the veracity of the Reverend Bell's claim. He arrived in his typical rural garments and convinced the officers, who immediately following the interview took him to an artist's studio where, for his contributions, a famous artist painted his portrait. The portrait can still be found in the Dorset County Museum. It shows a round, older man in country clothing, sitting a bit rigidly and showing his discomfort with the whole situation, as though he felt he did not deserve the fame. He was not like Dr. Jenner; he had done nothing extraordinary. He had only been protecting what was most precious to him: his family.

Story 10.3: A sick slave and the chain transfer of vaccines across the Atlantic

Hernán Cortés, whose full name was Hernán Cortés de Monroy y Pizarro Altamirano, landed on the eastern coast of what is today Mexico in early February 1519. With him were 630 men, eight women, 13 horses, and 12 arquebuses – a front-loading type of heavy gun. His "job" was conquistador. Cortés came to Mexico with one goal in mind: to conquer, colonize, and rule. He had no other plan; he had no alternatives. He was also an ardent gambler and was not afraid of taking risks, so he scuttled all the ships they had arrived on. It was a clear sign to the entire crew: no retreat. The only way was forward. About 200 miles away as the crow flies was Tenochtitlan, the capital of the Aztec Empire.

Cortés was not only a brutal conqueror, he was also a brilliant tactician. He made allies out of some of the native tribes that were under the rule of the Aztecs and gradually increased the size of the army under his leadership. However, despite his great determination and his strategic and tactical skills, his vision did not have much chance for success. He was facing a strong, centrally-organized empire that was at the pinnacle of its power at the time. In times of war it could deploy 700,000 troops.

But Hernán Cortés, and the conquistadors that came after him, would have an unexpected and cruel ally, one that killed more native inhabitants than all the conquistadors together could have ever done. It had not yet arrived, but it would be brought over by the ships that sailed to the shores of the Americas a year later; specifically, in April 1520.

Apart from soldiers, these ships also carried slaves, and our story concerns one slave in particular by the name of Francisco Eguía. Few recognize the name as it has been long forgotten, despite the fact that he had a huge impact – albeit indirectly – on the expansion of Western civilization in the Americas, more so than any other single person. By the time the ships anchored on the coast of America, Francisco Eguía had smallpox, the deadly disease that Europe had already known for many centuries and which had claimed hundreds of thousands of lives every year. However high the smallpox mortality rate may have been in Europe, it pales in comparison to the mortality rate on the American continent.

Francisco the slave was the primary source of the contagion and he infected the soldiers who marched on Tenochtitlan, who in turn infected the Aztecs and other native peoples. The disease spread like wildfire and by the summer of 1520, it had reached the halfway point between the coast and the capital. By September, it had hit the surrounding cities. In July, the Aztecs were still able to drive off Cortés and his army from Tenochtitlan, but by October the contagion had come to the capital, where it raged for 60 days. Among the first to die was Aztec ruler Cuitláhuac. The disease killed the majority of the army and about a quarter of the population and those who survived went hungry because there were not enough people to take care of supplies. Weakened by disease and hunger, the Aztec army could not hold back the attacks of the Spanish conquistadors. On Saturday, August 13, 1521, Hernán Cortés entered Tenochtitlan and claimed victory. Wherever they went, his soldiers had to step over the dead bodies of Aztecs.

The Aztecs gave smallpox the name *huey ahuizotl* (great rash), and the spread of the disease through the country had a devastating effect. One Franciscan monk described it this way: "*Indians... died in heaps, like bedbugs. In many places it happened that everyone in a house died, and as it was impossible to bury the great number of dead, they pulled down the houses over them, so that their homes became their tombs*." Smallpox was not the only disease brought by the conquistadors against which the indigenous population had no defense mechanism, and the consequences can be called nothing less than an extinction. According to Aztec tax records, the population of the empire in 1518 was 30 million. Just 50 years later, Spanish officials counted only three million.

The consequences of smallpox and other infectious diseases were equally devastating for the Inca Empire, if not worse. The Inca Empire was located in the northwestern part of South America. The smallpox epidemic came from the north before the arrival of the first colonizers in 1532, and the spread of the disease was exacerbated by the pride of this huge empire: its advanced road system. In just a few years, a large part of the population was wiped out, with some sources suggesting as much as 94 percent. The smallpox epidemic was the first, but it was not the only epidemic in the history of the empire. A typhus epidemic broke out in 1546, followed in 1558 by smallpox along with influenza. Another smallpox outbreak came in 1589, diphtheria in 1614, and four years later there was a measles outbreak.

Smallpox came to North America a little later; the first records are from 1633. It spread across the continent and affected nearly all groups of the indigenous peoples. The disease was no less devastating than it had been in South America. In the mid-18th century, smallpox killed 500,000 Native Americans, with 30 percent of the indigenous population on the west coast and a large portion of the population living in the central plains killed by the disease. Smallpox was not confined to just the native population. Boston was hit by several epidemics between 1636 and 1721 and the residents tried to run from the disease, thus spreading it all along the eastern seaboard. Records indicate that over 130,000 died of smallpox during the American Revolutionary War (1775–1783).

Before the introduction of inoculation, the only way to deal with an epidemic was quarantine. Isolating those exhibiting signs of disease from the rest of the population did not help the sick one iota, but it did prevent the further spread of the disease. When necessary, entire cities were quarantined and cut off from the rest of the world. By the mid-17th century, colonial America had established quarantine law to prevent the spread of smallpox epidemics. The law required all suspicious cases to be reported, houses where they had smallpox were to be marked with a red flag, the homes of the sick were to be guarded to prevent contact with neighbors, and those who sailed into the ports were subjected to mandatory quarantine.

The end of the 18th century grew near and with it came Edward Jenner's groundbreaking discovery – vaccination with the relatively harmless cowpox to prevent people from contracting the deadly smallpox. Europe, from where this deadly disease had found its way to the New World, would also be the source of assistance as well. The disease may have been unwanted and arrived by some sort of mistake, but the help was deliberate. It went down in history under the name *Balmis Expedition* and was one of the greatest philanthropic missions the world had ever seen. It was also the first international healthcare expedition.

It is estimated that 60 million people died of smallpox in Europe in the 18th century. The disease was not choosy: it killed monarchs and members of their families right along with commoners. The dead included Czar Peter II, King Louis XV, and Bavaria's Maximilian III Joseph. In the late 18th century, smallpox struck the family of Charles IV of Spain. The king's brother Gabriel, his sister-in-law Maria Ana Victoria, and one of his young daughters, Maria Teresa, all died of smallpox. Another daughter, Maria Luisa, and the king's wife Maria Louisa of Parma, were infected with smallpox but they survived. Affected by the tragedies in his family, Charles IV decided to do something about the disease. He first had all the remaining members of the royal family inoculated, and in 1798 he issued a royal edict ordering the inoculation of the entire population of Spain. When he heard about the work of Edward Jenner, he issued another edict making the cowpox vaccination available across all of Spain.

Two years later, a smallpox epidemic broke out in several Spanish colonies in South America. New Granada (now Colombia) was the hardest hit and also sent the biggest pleas for help. Realizing that his colonies were suffering a calamity of enormous scale, Charles IV decided to send the philanthropic expedition to help. The main objective was to vaccinate as much of the population as possible, and they would use the cowpox vaccine. In addition, the expedition was to teach the people in the colonies how to prepare the vaccine and form vaccination boards that would organize continuing campaigns. King Charles IV signed the edict on June 28, 1802 and the following year, on November 30, 1803, the corvette *Maria Pita* set sail from the Spanish port of La Coruña, heading for South America and the southern part of North America. The expedition was led by Francisco Javier de Balmis.

At the time of the historic journey, Dr. Balmis was 50 years old and a zealous proponent of vaccination. He was also an experienced physician and traveler. He was born on December 2, 1753 in Alicante into a family where medicine was a tradition. He went to Mexico at the age of 25, where he would later become the chief surgeon at a hospital in Mexico City. The corvette's crew included the doctor's deputy director, two surgeons, five assistants, and 22 orphans aged between eight and 12. The only woman on board was the director of the La Coruña Orphanage, and she was charged with looking after the children.

The structure of the adult members of the crew is clear: this was a medical expedition of a scope never before seen, and that was why they needed both physicians and their assistants. But why the children? The answer is in the journey duration and conditions. At the time of the expedition, it took roughly 7–12 weeks to sail from Europe to the Americas, depending on the conditions. Since there were no refrigerators or freezers, a delicate vaccine like that may not have survived such a long journey. One option was to transfer the cowpox vaccine in a human who was inoculated in Europe and who would carry the vaccine to the Americas, but the problem was that during the trip, the vaccine carrier would have plenty of time to recover from the disease and would only end up bringing his immunity to the new place. So how could they bring the vaccine across the Atlantic Ocean and keep it viable? The solution was a transmission chain. During the journey they would keep inoculating healthy volunteers, one by one, so that when they arrived in the Americas the last to have been inoculated would be bringing the "living" vaccine. But where would they find volunteers? The answer was found in the 22 orphans from the La Coruña Orphanage. They vaccinated the boys in pairs and when one pair of boys developed the infectious pustules of cowpox, another pair was infected with them. The pattern was repeated for the entire journey.

By today's standards, the transmission of pustule fluid from one child to another in order to create a chain transmission may not be acceptable, but it was an effective and creative method at a time when refrigeration, sterile containers, and asepsis had not yet been invented.

On February 9, 1804, ten weeks after they set sail, the expedition landed in Puerto Rico. They established the first vaccination board and continued southward to what is now Venezuela. They were met with great fanfare in Caracas, and after establishing another vaccination board, Balmis decided to split the expedition into two groups. One group was led by his deputy director and they traveled south, while Balmis and his group went north. They vaccinated 12,000 people in Venezuela. On June 25 that year, Balmis went to Mexico via Cuba.

His deputy, Dr. José Salvany, encountered substantially more difficulties on his travels to the south. They were shipwrecked on their way to Colombia and the crew barely managed to save their precious cargo. As soon as they landed, however, they began with the vaccinations. When they were finished on the coast, they set off on a difficult journey inland to Bogota and to make sure they would bring a living vaccine to their destination, they took ten young boys from a local orphanage who would carry the vaccination material. They traveled up the river and at each port, they continued the vaccinations. They vaccinated a total of 56,000 people. Along the way, Salvany became seriously ill and lost an eye. They arrived in Bogota several months later, on December 18, where they were warmly welcomed. Upon finishing their work in Bogota, they headed to Quito (Ecuador) where they gathered more young boys, and then crossed the Andes into Peru. At every stop they vaccinated thousands and thousands of people, nearly 20,000 on the way from Quito to Lima. From Peru, they continued on to Chile and from there to Bolivia. On January 28, 1810, Salvany died at Cochabamba, Bolivia.

Francisco Javier de Balmis did not continue the vaccinations in the Americas. In February 1805, he sailed westward from Mexico across the Pacific Ocean. By royal decree, he vaccinated people in another Spanish colony in the Philippines; at the time he was accompanied by 25 Mexican orphans. After stopping in Manila and in Canton, China, he completed his two-and-a-half-year journey around the world in July 1806.

The Balmis – or probably more accurately the Balmis-Salvany – expedition was the first official mass vaccination program in Latin America. It was an important medical campaign not only for how extensive it was, but also because, under the auspices of King Charles IV, a new medical technology was institutionalized for the first time. It was also a visionary activity of a kind that had never been seen before. Edward Jenner himself commented: "*I don't imagine the annals of history furnish an example of philanthropy so noble, so extensive as this.*"

Story 10.4: Two greats and only one Nobel Prize

Edward Anthony Jenner developed, described, and promoted the first vaccine in 1798. It originated from cowpox and it caused a person develop immunity to its deadly relative, smallpox. Almost immediately, it spread around the globe, and by the early 19th century, with the help of the 22 orphans, it had sailed across the Atlantic and landed in the New World. But humanity would have to wait 100 years, until the late 19th century, for vaccines against other diseases.

The reason was simple. The smallpox vaccine was developed solely on the basis of observation, illustrating how even pure empiricism, experience, and a little courage could lead humankind to a revolutionary discovery. Even windmills and watermills, the steam engine, and many other inventions were conceived only on the basis of observation, long before scientists described the principle of how they functioned.

When it came to other diseases, there were no such visible links between the disease itself and immunity to it. Further knowledge would be necessary, and not just knowledge; someone would also have to come up with a coherent theory that would result in practical knowledge that could be applied, for instance, to the development of additional vaccines. In this case, a new theory was not the only thing that was formulated. A whole new science was invented: bacteriology. The creation of bacteriology is closely associated with two famous scientists – Louis Pasteur and Robert Koch.

Although their greatness was not based on the discovery of new medicines, these two titans of science cannot be left out of a book about the history of drugs. They uncovered something even more important, something without which drugs could not be invented. They discovered and described the causes of most of the diseases known at the time.

Louis Pasteur, the older of the two, opened the pathway to understanding that diseases do not occur spontaneously or as the result of some higher power, but are instead caused by living microorganisms that have the ability to multiply. He made his discovery not while studying diseases, but while attempting to prevent wine from souring. Pasteur discovered that the alcohol in wine did not occur spontaneously but that it was created by microorganisms, specifically yeast. He also discovered that different microorganisms – bacteria – continued the process and turned the alcohol acidic, causing the wine to sour.

Pasteur was the first to create a vaccine against a bacterial disease. The smallpox against which Jesty and Jenner used vaccination as protection was caused by a virus. Pasteur's vaccine remains a first, even though it was not used to protect humans but to protect chickens, and even though, like many discoverers before and after him, help came in the form of chance. At least, according to the legend.

In the 1880s, Louis Pasteur was working on a poultry disease known as chicken cholera. He identified the cause of the fatal disease as bacteria and was then able to isolate the bacteria and grow them in culture media. The story goes that Pasteur was getting ready to go on vacation in 1880, but before he left he instructed one of his assistants to inoculate healthy chickens with a culture of the isolated bacteria. However, the assistant forgot and then went on vacation too. When he returned, he found the month-old culture plates and remembered what he had been instructed to do. He decided to correct his mistake and inoculated the chickens with the old culture. The chickens showed signs of disease, but not one of them died, and in time they all recovered. When they were later infected with a fresh bacterial culture, the chickens did not get chicken cholera. They were immune to it.

That brings us to the third method by which vaccines are prepared. The first, and very risky, method was brought to Europe by the beautiful and stubborn Mary Wortley Montagu in the early 18th century. It worked on the principle that the application of viral smallpox into the body via a small scratch on the arm was less harmful to the carrier than droplet infection, which was the primary way by which this fatal disease was transmitted. The symptoms of the disease were milder, but the body could develop immunity even despite the milder case and the person would never get smallpox again. The second method is that of Benjamin Jesty and Edward Jenner. They both used the empirical evidence which showed that infecting the body with a disease that is less deadly to humans creates immunity to other, far more deadly diseases related to the less deadly one. An infection with cowpox caused immunity to smallpox to develop.

The third, and most frequent, form of immunization is the application of the weakened disease agent against which immunity is desired. A weakened disease agent, such as the chicken cholera bacteria that were left on a laboratory table for a month, is not as harmful as an unweakened disease agent. The important thing is that despite its being weakened, it retains the ability to create defense mechanisms in the body to protect from the given disease. Technically, it is the reduction of the virulence of the causative agent without reducing its immunization abilities.

Only a few years after Louis Pasteur announced his discovery to the medical community, a humanitarian appeared who would use it to protect people. His name has been long forgotten, so let us reintroduce him. He was Jaume Ferran i Clua, a Spanish physician and bacteriologist. By simple analogy, he assumed that if the weakened bacteria could be used to create immunity in chickens, the same thing would work in people too. He therefore prepared a vaccine for human cholera using a method similar to Pasteur's, and during the epidemic in Valencia, Spain, he first tested it on himself and then vaccinated over 30,000 people. That was in 1885. Cholera thus became the second disease, after smallpox, against which a vaccination offered protection.

Louis Pasteur's successes did not end with a vaccine for chicken cholera. He may have arrived at the principle of vaccination by accident, but he put the knowledge to good use. He decided to use it to prepare a vaccine for another dangerous and deadly disease known as anthrax. This was a disease that afflicted cattle and it caused huge damage to farmers, so if it could be eradicated, it would be of great economic benefit. Pasteur prepared the vaccine by weakening the live, deadly anthrax bacteria by heating them to 35 °C. He first tested the vaccination in laboratory conditions, and when he ascertained that it worked, he was ready to take on anthrax.

At this time, however, Louis Pasteur's discoveries were still thought to be extremely revolutionary and the French scientist still had as many opponents as he had supporters. Those opponents demanded that he prove the vaccine's effectiveness with a public experiment. Pasteur was given two herds of livestock, each consisting of 25 sheep, one goat, and several cows. One herd was vaccinated, the other remained unimmunized. He vaccinated the one herd twice, on the first day and then on the 15th day. He waited another 15 days, and then he was ready for the final step. The date of May 31, 1881 and the exact location of the experiment – a farm at Pouilly-le-Fort – have both gone down in history. The demonstration was attended by scientists, journalists, and over 100 farmers. That morning, Pasteur's assistant inoculated both herds with live, virulent anthrax bacilli and everyone waited anxiously for two days for the results. The experiment was an absolute victory for Louis Pasteur. All the animals in the control group died, and all the vaccinated animals survived. The public response was overwhelming and soon Pasteur's vaccine spread to the rest of the world.

In 1879, Louis Pasteur abandoned fowl and livestock diseases and began to work on a disease that, compared to the great epidemic diseases, was not very common: rabies. Long before Pasteur came around, it was known that rabies was transmitted to humans through a bite by a rabid animal, usually a dog, and that once a person contracted rabies, it was virtually always fatal. In one of his papers, Pasteur recorded the following idea, characterizing not only his manner of thought but also the considerations of those who came after him: "*If rabies could be attributed to the action of a microscopic organism, it would perhaps no longer be beyond the natural resources of science to find a means of attenuating the action of the virus of this fearful disease, and thereafter put it to use, first to protect dogs and then to protect humans*."

Louis Pasteur and his team worked for six years to attenuate the action of the rabies virus. On Monday, July 6, 1885, Pasteur administered the rabies vaccine to nine-year-old Joseph Meister, who had recently been bitten by a rabid dog. It was a success. The boy developed an immunity that saved him from certain death. It was risky for Pasteur, as he was not a practical doctor and was thus not authorized to provide treatment to anyone, but young Joseph recovered and Pasteur's infringement was forgiven. The boy was also the first human in whom a vaccine was used that was not in prevention of a disease. The purpose of the vaccination was to prevent the existing disease from developing any further. This is known as post-exposure prophylaxis. As well as rabies, it is now also used with other diseases, including HIV.

Louis Pasteur was a science heavyweight. He towered over everyone who came before him and those who came after him. There is probably just one other scientist whose name is spoken with the same reverence as that of this French chemist and microbiologist. That would be his contemporary, the German physician and microbiologist, Robert Koch.

There are common behavior traits that we ascribe to different nationalities. Louis Pasteur was a true Frenchman whose behavior strictly adhered to the preconceived notions we have of all French people. He was bold and inventive, and not very systematic. He was a passionate explorer who constantly thought up correct theories and incorrect estimations. He jumped from one problem to another and his conclusions were not always substantiated with sufficient evidence. His zealous defense of his own results once even led to a physical altercation, at the age of 60, with an opponent who was 20 years older. The two seniors would have engaged in a downright brawl in front of their colleagues had the younger and stronger ones not pulled them apart.

Heinrich Hermann Robert Koch was German. Actually, he was Prussian. Coldly logical, like a geometry textbook, his results came as the logical consequence of his experiments and he defended his findings from a distance, as though they belonged to someone else. He substantiated his successful experiments with bullet-proof evidence, and he approached his failures with the same resolve. When he wanted to test the infectiousness of the tuberculosis bacilli that he had first isolated and then cultivated, he filled his laboratory with a whole array of animals. He did not confine himself to the standard laboratory mice, rats, guinea pigs, rabbits, and monkeys; he also had turtles, sparrows, five frogs, three eels, and one goldfish.

First, Koch discovered the bacteria responsible for anthrax. Anthrax already came up in this story when we talked about Louis Pasteur developing a vaccine, but Pasteur did not do any in-depth studies of the microbe. What he did, in a nutshell, was transfer infected material from one animal to another, or from an infected animal to a culture medium and from there to a healthy animal. When the healthy animal became sick, it was evident that the microbe was pathogenic. Koch, however, studied the anthrax bacteria. He monitored their growth and transferred them from one organism to another. He isolated them in a culture medium in a way that left only these bacteria and nothing else that could cast doubt on his hypothesis that it was these microbes that caused the disease. He was therefore the first to link one specific microbe with one disease and he did the same when studying other diseases. He was the first to see, isolate, and describe the bacteria that cause cholera and tuberculosis. In fact, it was the discovery of *Mycobacterium tuberculosis* – the microbe that causes tuberculosis – that immortalized him, and it is little wonder. In the late 19th century, the death of every seventh German was caused by the disease.

Unlike Louis Pasteur, Robert Koch brought organization and order to microbiology. He worked up precise rules about when and under what circumstances a causative agent could be linked with the disease itself. His rules are still used in microbiology today, and they are so much a part of the standard practices that few probably ever think about how this Prussian microbiologist invented them 125 years ago.

There is one more reason why this scientific heavy-hitter has a place in this story: his name is associated with a particular vaccine. Although, as Koch himself admitted, it was a vaccine that was more than controversial.

Robert Koch announced his discovery of the causative agent of tuberculosis to the science world on March 24, 1882. Just over eight years later, on August 4, 1890, he announced something even more impressive. He had discovered a drug to treat tuberculosis, a vaccine called tuberculin. He did not want to disclose the composition of the drug or how it was prepared, but the response was overwhelming. The researcher was a recognized expert so there could be no doubt as to the significance of his discovery. Physicians praised him for his triumph over the most voracious disease of the time, while tuberculosis patients from all over the world began to converge on Berlin. Everyone wanted to get their hands on the new drug as fast as they could. An entire 150-bed ward was designated for the study of tuberculin in order to facilitate the first results of the new treatment. The director of the ward was Paul Ehrlich, the discoverer of Salvarsan, a drug that was the subject of a story in the penicillin chapter.

Soon, however, articles began to appear in medical journals about the first problems. There was no such thing as clinical trials at the time tuberculin was discovered, and experience with the action of the drug on laboratory animals was applied directly to human therapy. The problem is that laboratory animals cannot complain about adverse reactions like headache, fever, and nausea and it appears that these adverse reactions were more frequent and more intense in humans than they were in the animals. There was even news about deaths associated with the drug, and Robert Koch was forced to divulge the composition of tuberculin. It turned out to be a glycerin extract of the bacteria, but that was not the worst of it. It was discovered that tuberculin had practically no effect and it thus gradually faded away until it was all but gone for good. Until 1907, that is, when a scientist who came after Koch began using it as a diagnostic tool for tuberculosis. It is still used as a diagnostic tool today.

As a drug, tuberculin has long been forgotten, fading into obscurity and leaving behind only a lesson on how not to introduce a drug into clinical practice. Likewise forgotten is Louis Pasteur's experiment, which today would be considered completely unethical – a non-doctor daring to treat a human being, and a young boy at that. All that is left are the results and the names of two great men. Most historians place them alongside one another on the same level. Interestingly, only one of them would win the Nobel Prize: Robert Koch, in 1905. Louis Pasteur did not win the Nobel Prize because it did not exist until after the death of Alfred Nobel, who died in 1896, and it was awarded for the first time in 1901. The prize was only awarded to laureates living at the time of the award and Louis Pasteur died in 1895, before Alfred Nobel. Had he lived a few years longer, Pasteur most certainly would have received the Nobel Prize. It would not have been awarded only to his Prussian rival.

Story 10.5: "Sir Almost Wright" and military brains

Even today, hearing the words typhus or typhoid makes people shiver, despite the fact that these diseases have been practically eradicated in Western civilization. Typhus and typhoid are two different diseases whose names derive from the Greek *tûphos*, meaning, among other things, hazy. The diseases were named for one of their symptoms, a stupor or hazy mind.

This story is about typhoid fever, the second of the human infectious diseases after smallpox for which a vaccine was developed and used on a large scale. The development of this vaccine is linked to one of the greatest bacteriologists and immunologists of the late 19th and early 20th centuries. His name was Almroth Edward Wright, and he was not only a distinguished scientist but also a man of contradictions. Before we delve into this unusual and colorful character in the history of medicine, let us briefly look at the actual disease.

The symptoms of typhoid fever are similar to epidemic typhus, with fever, headache and stomach ache, and colored spots. The main difference between the two is that typhoid fever is caused by a different bacterium than typhus and no special carrier is needed to transmit it to others. The bacterium is a type of *salmonella* and easily spreads due to poor hygiene.

The first known epidemic attributed to typhoid fever broke out in AD 430 in Athens and killed nearly a third of the population of this city-state. It changed the history of the ancient world, as it resulted in the end of the dominance of Athens and shifted the seat of power to neighboring Sparta. But the 19th century was the golden age of typhoid fever, and in fact of almost all other infectious diseases. Industrialization, urban population growth, and densely populated and squalid slums with poor hygiene all contributed to the spread of epidemics. Take for instance the long, hot summer of 1858 in London. The fetid odor of the waste that flowed unchecked into the River Thames made life in the metropolis practically impossible. People were suffocating and vomiting, and sessions of the House of Commons had to be suspended. This period went down in the history of London as the Great Stink. Other large cities faced a similar situation.

Typhoid fever was also a serious threat in times of war, when epidemics greatly reduced the ability of the armed forces to engage in battle. The first measures to improve conditions came from within the military, and soon such measures were undertaken in civilian areas. Those first steps were improvements in hygiene, and vaccinations. Now we can meet Almroth Wright.

He was born on August 10, 1861, near Richmond in the north of England. His father was the Reverend Charles Henry Hamilton Wright, an Irishman and the leader of the Protestant Reformation Society. His mother, Ebba Almroth, was the daughter of the governor of the Swedish Royal Mint in Stockholm. His background greatly shaped his personality. Almroth Wright was charismatic, able to inspire absolute loyalty in some and great aversion in others. His detractors played word games with his name, calling him "Sir Almost Wright" and even "Sir Always Wrong". He was tall, clumsy, had a hunched back from leaning over his laboratory table, and moved with all the grace of a toddler, but he was also dexterous and precise when it came time to exacting laboratory procedures. He was kind and gentle with patients, yet openly hostile towards his colleagues. He was a charming

Celt and a great storyteller. He was a fixture of London society and a personal friend of George Bernard Shaw.

At the age of 21, Wright graduated from Trinity College in Dublin with a degree in modern literature. He simultaneously took courses in medicine, graduating a year later. He had a phenomenal memory and it was said he could recite a quarter of a million poetry lines by heart. We were already supposed to have come across Almroth Wright in this book, in the chapter about penicillin, as he was in charge of the Inoculation Department at St Mary's Hospital in London, where young Alexander "Sandy" Fleming took a job after he finished school. It was because of Wright that the young graduate became a bacteriologist who later gifted humankind with penicillin. Ultimately, though, Wright was cut from that chapter because he did not recognize the process advocated by Fleming and other bacteriologists. He believed it was pointless to search for molecules that could be targeted to fight infectious diseases. The only way that he saw that made any sense was vaccination. He was Mr. Almost Wright.

Wright began working on a typhoid vaccine in 1893, just after becoming professor of pathology for the Royal Army Medical Corps at the Royal Victoria Hospital in Southampton. As already mentioned, typhoid was more dangerous for soldiers than enemy projectiles. Encouraged by Louis Pasteur's experiments ten years earlier with the anthrax vaccine, Wright decided to use a similar method to prepare a typhoid vaccine. However, he faced certain challenges that Pasteur had not. Pasteur had used weakened, albeit live bacteria to induce an immune reaction, but notably, the Frenchman was able to prove the effectiveness of the vaccine by inoculating animals with the live, virulent bacteria of the disease itself. It was fine to conduct this experiment on animals, but it was a completely different matter to do so on humans.

Wright overcame the first hurdle by using dead bacteria. They could not actually cause the disease itself, but they had the ability to induce the production of antibodies that protect humans from becoming infected with live bacteria. This remains one of the most common methods of vaccination preparation. Next, he developed a simple method that would allow him to determine, after the vaccine was given, whether antibodies against the disease were forming in the serum of the vaccinated individual, and if so, the quantity. He determined the production of the antibodies *in vitro*. It was no longer necessary to administer live (and dangerous) microbes to humans to ascertain the effectiveness of a vaccine.

As soon as he had an effective vaccine, Wright used it. He first injected himself and then 16 devoted coworkers. The first field test involved administering the typhoid vaccine to 2,835 British soldiers who were on their way to India. Despite the high incidence of typhoid in India, only six of the soldiers contracted the disease and Wright took that as definitive proof that the vaccine worked. It was the time of the Boer Wars in South Africa, where the incidence of typhoid was extremely high, and Wright proposed to the War Office that all soldiers heading to the South African front should be inoculated. The War Office did not share Wright's enthusiasm, however, and demanded more proof with a greater number of vaccinated soldiers. Wright argued that evidence would only be obtained if the army allowed the inoculations. This created a vicious circle, and it was in this context that the inventor uttered his now famous reply: "*No sir, I have given you the facts. I can't give you the brains.*" Wright did not prevail, however, and just a small percentage of the soldiers sent to South Africa were vaccinated. During the three years of war, 59,000 British soldiers came down with typhoid, of which 9,000 died.

Many research scientists would probably have given up after a disappointment like that. Injecting microbes, whether dead or nearly dead, into human beings must have seemed absolutely disgusting to the officers in charge. But Almroth Wright was half Irish (and half Swedish) and he persisted. He continued trying to persuade those in high places, had his results re-examined by another (less combative) colleague, and gradually began winning people over. Ultimately, the War Office accepted the vaccination and in 1915 the vast majority of the British troops were inoculated against typhoid. The outcome was that for the first time, the number of soldiers killed in battle was higher than the number that died of an infectious disease. That may seem like a hollow victory, but as far as vaccinations are concerned, it was a success.

The typhoid vaccine was far from the last contribution made by Almroth Wright for the benefit of humanity. In addition to the vaccine to prevent typhoid, he also developed a vaccine to treat typhus and he was instrumental in the development of both a tuberculosis vaccine and a pneumococcal vaccine.

Finally, a few words about how accurate his nickname "Sir Almost Wright" was. Despite the unquestionable evidence, he did not believe that a vitamin C deficiency was the cause of scurvy. He held that the disease was caused by rotten meats. He also claimed that the brains of women were completely different from those of men and therefore, women should not hold positions in social and public affairs. He was opposed to women having the right to vote and he refused to hire women for his research team. Sir Almroth (Almost) Wright.

Story 10.6: The Righteous Among the Nations and lice feeders

Probably the most famous outbreak of epidemic typhus in history occurred in 1812, when Napoleon's massive army headed east to conquer Russia. They were a force to be reckoned with. On June 22, 1812, an army of 680,000 determined and experienced warriors crossed the border at the Neman River, at the point that now forms part of the Belarus–Lithuania border. Less than three months later, Napoleon arrived in a deserted Moscow with fewer than 100,000 soldiers, most of whom were weak and suffering from malnutrition. Those who did not make it to the destination had stayed back in makeshift barracks, sick and weak, dying one after another. Mostly of typhus.

Pyotr Ilyich Tchaikovsky composed one of his most famous pieces, the *1812 Overture*, to honor the Russian victory over Napoleon and there is no doubt that the huge sacrifices made by the Russian people are the main reason for Napoleon's failed invasion of Russia. But a few notes of this monumental work could rightfully be dedicated to a microscopic participant in the campaign: the body louse, or *Pediculus humanus* in Latin. Without it, Napoleon's dream of a magnificent French empire, reaching from Russia all the way to India, may not have been so definitively crushed.

Epidemic typhus – the subject of this story – is caused by the bacterium *Rickettsia prowazekii*. It was named after the two scientists who discovered its role in the spread of the disease, American pathologist Howard Taylor Ricketts and Czech zoologist Stanislaus von Prowazek. Research into the causative agent of typhus was not without its victims, and both of these men died while studying the disease. *Rickettsia* is transmitted from victim to victim by the body louse, a small insect that lives in clothes, favoring wool and cotton. When the louse is hungry, it crawls out of the clothing and attaches itself to the skin to feed on blood. If the blood "donor" is infected with typhus, the louse also becomes infected and after a time it will die. However, if in the meantime it is transferred to another person, its remnants and feces can easily get into small cuts and scratches and transmit the typhus to that person. It can also be transmitted by breathing in dried louse feces found in clothing or bedding. Imagine all those soldiers living day in and day out in the same uniforms, crowded and pressed up against one another at night, and it is not hard to see why in some languages war was given the unflattering attribute *lousy*.

Lousy wars did not end with Napoleon's withdrawal from Russia. During World War I, 105,000 people died in Serbia alone and it was not just the army that was lice-infested, civilians were too. Typhus was one of the main causes of death during the Great Famine in Ireland in 1846–1849, while over 25 million people died of typhus in the Soviet Union in 1922−1923. We could go on and on, but the end was nigh for this life-threatening disease. It would be brought down by the unassuming and quite unknown Polish biologist, Rudolf Weigl.

Rudolf Stefan Weigl was born on September 2, 1883 into an Austrian family in Prerau, Moravia. Little is known about his father, but he purportedly worked for a transportation company and died while testing a new bicycle. Now a widow with three children, Rudolf being the youngest, his mother went on to marry a Polish school teacher. Together, they moved to the Polish-speaking Lvov, and Rudolf gradually adopted his new home. He was a Pole, despite the fact that all residents of Lvov lived at one time or another in the Austro-Hungarian Empire, then in Poland, then in the Soviet Union for a while, then in Germany, and then again in the Soviet Union.

Weigl attended the University of Jan Kazimierz in Lvov, and after graduation he worked as a parasitologist at the army parasitology laboratory. This is about the time he became interested in combatting typhus. When he became a professor of biology in 1920, he brought that interest with him to the university. There, he would build one of the best-known laboratories in Europe for the study of *Rickettsia*. Some would have been satisfied with that, but Rudolf Weigl decided he would not only study the microbes but he would also use them to develop a vaccine for epidemic typhus.

It was already known at the time that the causative agent was *Rickettsia* and that it was transmitted from person to person by the body louse. The problem was that *Rickettsia* could not be cultivated anywhere outside the stomach of a louse. That meant that if Rudolf Weigl wanted to make a vaccine against epidemic typhus, he would have to be the first person in the world to tame lice and turn them into laboratory animals. Not only that, body lice can only survive if they feed on human blood, so human feeders had to be assigned to the laboratory lice.

The technique developed by Rudolf Weigl to prepare a vaccine against epidemic typhus was really quite amazing. In fact, the entire life story of this modest scientist is something that just cannot be made up. His story is not only a part of the history of medicine, it is also a powerful lesson in humanity.

His technique began with lice laying eggs on a small piece of wool fabric. He incubated the eggs in a glass vessel, and they hatched at a temperature of 90 °F. Once they hatched, the larvae were placed into the first of Rudolf Weigl's inventions – small, wood cages measuring 4 × 7 cm and only 5 mm deep. The back wall of the cage was fitted with a special screen that had originally been used to sift flour. The screen allowed the louse larvae to stick their heads through but they could not escape. About 800 larvae would be placed into one cage, and the cages were then sealed with paraffin to keep the lice from escaping. The larvae were now in place and it was time for the next step, the feeding.

Depending on the size of the feeders' thighs or calves, 7–11 larvae cages were attached to their legs. They were fixed with elastic bands, with the screen side of the cage pressed against the skin. The larvae then stuck their heads through the screen and fed on the human blood for about 45 minutes. Afterwards, the cages were removed to prevent the larvae from overeating. The feeding usually lasted for 12 days and the red bitemarks left after each feeding were disinfected with alcohol. Men usually had the cages attached to their calves, while women had them attached to their thighs so they could conceal the red marks beneath a skirt. Most of the feeders handled the minor discomfort, and the minor blood loss, quite well.

The story is already amazing, but there is more. Entire colonies of healthy lice were of no use – they had to be infected with the *Rickettsia* bacteria. This is where Rudolf Weigl's second invention came into play; small clamps that were constructed so that larvae could be safely secured in place but not crushed. They were secured with their rear ends in the air and a small glass pipette containing the required amount of *Rickettsia* was inserted into the anus of each louse under a microscope by trained injectors. A team of two experienced operators could infect up to 2,000 lice an hour.

But there was still more to it. The infected lice had to feed for another five days and this is where a problem came up, because the lice were now infected with the microbe that causes typhus. All the feeders would need a strong vaccination to keep them from getting the disease. Some of them actually did contract typhus, but fortunately, no one died.

The final step was strictly a laboratory procedure. The lice were killed and their guts were removed under a microscope, ground into a paste, and put through a centrifuge. The result was a solution that could be diluted according to the desired strength of the vaccine, then the vaccine was sealed into glass ampules and ready for distribution.

That was how the vaccine against epidemic typhus was produced in Rudolf Weigl's laboratory in Lvov. Production continued for less than ten years, from the mid-1930s until Lvov was occupied by the Soviet Union in 1944.

Professor Weigl's laboratory lice method is no longer used to produce typhus vaccines. It has been replaced with new, more effective and more humane methods. We are left with one essential question: who were these feeders, these people who would let themselves be bitten by lice – even lice that were infected with epidemic typhus?

At first, it was most likely employees of Professor Weigl's laboratory, including Weigl himself and his wife, son, brother, and cousins. The first turning point occurred in 1939, when the Soviet Union occupied Lvov under an agreement with Germany. At the time, Lvov was Poland's second largest cultural and academic center after Warsaw. Poles comprised two-thirds of the population; the rest were Ukrainians and Jews. The occupation was followed by intense Sovietization. The Poles were deported to the Kazakhstan plains and many men ended up in Soviet gulags. To a large extent, the oppressed were Polish intellectuals. Although Professor Weigl was an Austrian from Moravia, he decided to help the Poles by hiring them as feeders. The typhus vaccine was a strategic material, so the Soviet officials let this little violation slide. Slowly, professors of math, sociology, history, and geography, as well as writers, musicians, and more academicians and intellectuals, all became feeders. Had a person peeked into Professor Weigl's laboratory, they would have seen groups of Polish intellectuals as they sat in a circle discussing math or history – with cages full of lice attached to their legs.

The situation worsened in 1941 when the German army occupied Lvov, but the Germans needed the typhus vaccine too. The laboratory in Lvov was officially attached to the military Institute for the Study of Epidemic Typhus and Viruses *(Institut fur Fleckfieber und Virusforschung)*, and Rudolf Weigl had a bit more leeway. Vaccine production had to be expanded constantly, so he was able to hire more and more feeders. His employees were given their own identification cards that allowed them additional food rations, as well as a relatively significant amount of freedom. No patrols wanted to get up close and personal with people who were half-eaten by lice and might be carriers of typhus.

But Polish intellectuals were not the only ones to find sanctuary at Professor Weigl's institute. Members of the Polish resistance movement (*Armii Krajowej*) also came, as it was important for them to hide out during their activities. When the Germans began the systematic murder of the Lvov Jews, many of them also became feeders. Ultimately, over 4,000 people passed through Professor Weigl's institute. For most of them, participation as lice feeders probably saved their lives.

Saving Lvov's Jews by employing them as feeders was not Professor Weigl's only contribution that later earned him wide-spread recognition. Members of the Polish underground resistance movement smuggled some of his vaccines to the Lvov and Warsaw Ghettos and some even found their way to the concentration camps at Theresienstadt and Majdanek.

In 2003, nearly 50 years after his death, Professor Rudolf Weigl was granted the title of Righteous Among the Nations, a recognition awarded annually to non-Jews who helped save Jews during the Holocaust. Rightfully so.

Story 10.7: The Somali cook and a huge victory

Let us return for a minute to the first story in this chapter, the one about the beautiful Lady Mary Montagu. We started off with a pretty detailed description of the symptoms of smallpox but the entire description and all the symptoms were given in the past tense. The reason is simple. It would be difficult to describe a disease that no longer exists in the present tense and since it does not exist, we can only mention that it once did. Past tense.

The last to person to be infected with smallpox was Somali cook Ali Maow Maalin. It happened in October 1977 and Ali made a full recovery. After that, not one more case of smallpox was ever recorded. Everyone waited for two years for another "Ali" to turn up, but it did not happen, and on October 26, 1979, the World Health Organization (WHO) declared that smallpox, one of the greatest blights in the history of humankind, was a threat no more. The disease had been eradicated.

Edward Jenner, inventor of the first vaccine, had dreamed of the possibility of eradicating smallpox. In 1801, five years after he first applied the vaccine to James Phipps, he wrote "*It now becomes too manifest to admit of controversy, that the annihilation of the Small Pox, the most dreadful scourge of the human species, must be the final result of this practice*."

It was only a vision at the time, one that would take another 178 years to be realized, but smallpox did warrant such a vision. The fight against smallpox had several advantages compared with many other epidemic diseases, the first of which was that the disease was clearly distinguishable. The skin rash was visible and characteristic of the disease. Also, there was a relatively short amount of time between infection and the appearance of the first symptoms, so it was not too difficult to track down all the people the patient had come into contact with. Another advantage was that smallpox could only be transferred from human to human. There were no other possible carriers, like rats or fleas with the plague, mosquitoes with malaria, or the lice that transmit epidemic typhus. There was also the fact that the smallpox vaccine was highly effective, which meant that large groups of people could all be immunized in a fairly short amount of time.

The only thing that was required was to vaccinate everyone who had contracted the disease and everyone who had come into contact with it. It seemed easy enough. Except that when the WHO accepted its first proposal to eradicate smallpox in 1959, there were three billion people in the world. One-third of all the people lived in endemic countries, where the disease was constantly maintained among the population without external sources of the infection. While it is true that smallpox had been eradicated in Europe and North America by that time, even the countries situated in these zones could not feel completely safe. All it would take was one infected traveler and an epidemic could break out just about anywhere.

Take for example the last European epidemic, which broke out in the former Yugoslavia in 1972 when a Muslim pilgrim named Ibrahim Hoti brought smallpox to his village in Kosovo. He had spent a few days in Iraq on his way from Mecca and had probably become infected with smallpox there. He returned home in mid-February 1972 and the first victim was a 30-year-old school teacher who died in early March. There were soon 140 cases of smallpox in Kosovo and a panic broke out. Martial law was declared in the entire area, there were roadblocks, all social events were cancelled, and a travel ban was put in place. Hotels were commandeered to quarantine the 10,000 people who came into contact, or may have come into contact, with infected persons. With the help of the WHO, nearly the entire population of Yugoslavia was revaccinated, almost 18 million people. These drastic measures were successful in containing the epidemic and only 175 people contracted the disease, of which 35 died.

Ten years earlier, in the 1960s, the eradication initiative was very slow getting off the ground. That was not actually surprising, since vaccinations were necessary not only in cities and villages, but also in hard-to-reach places such as in the Amazon and among nomadic African tribes, in the marshes of Bangladesh and high up in the Himalayas.

Things started to improve in 1967, when the World Health Assembly approved a special budget for the project. That the members themselves were ambivalent about the entire exercise is evident in how they voted: the budget was approved by just two votes above the minimum required. It was the narrowest margin in the history of the WHO.

The second crucial factor was the engagement of Donald A. Henderson to lead the campaign. With a small team of four people, he was able to coordinate the work of dozens of local advisors who were responsible for vaccination teams in their areas. He had to deal with the huge distances between the epicenters of the disease, as well as bureaucratic obstacles, resistance caused by religious and social bias, and the dangers of armed conflicts. Smallpox was still endemic in dozens of countries at the time, with 15 million cases of the disease annually and over two million deaths.

The campaign began in Brazil. The population of that country was 96 million, a large part of which lived in remote Amazon rain forests. By then, several South American countries had declared the eradication of smallpox and Brazil was the last refuge for the disease. Mass vaccinations of all the people in this enormous country were impossible, so instead they used a model that was a combination of surveillance followed by containment. The minute the local team had information about a breakout, they immediately traveled to the location, hermetically sealed the entire area, and vaccinated all those who were sick and anyone who may have come into contact with them. Just like in Yugoslavia in 1972. Between 1969 and 1971, 63 million people were vaccinated. The campaign was successful, and in 1971 smallpox was eradicated in Brazil. South America was now a continent free of smallpox. Ironically, the last cases of the disease did not occur somewhere in the Amazon jungle, but in the capital city of Rio de Janeiro, just a few hundred yards from the campaign's headquarters.

Indonesia was another problem area. At the time it had a population of 120 million, spread across 3,000 islands. Only 60 percent of the populated areas could be reached by car, 20 percent could only be reached by boat, and the remaining 20 percent could only be reached on foot or by bicycle. There were half a million cases of smallpox. The eradication program began in 1969 in Java, and bus drivers, soldiers and businessmen were recruited to search for cases. The goal was to register outbreak sites as soon as possible so that medical teams could be sent. Three years later, they were able to announce that there was no more smallpox in Indonesia.

With a population of 550 million, India was the largest of the endemic countries, with half of all the reported cases of smallpox found there. There were temples to Shitala Mata, the goddess of smallpox, throughout the country. Smallpox had been present, and prevalent, in India for so long that many no longer believed it could be eradicated from their lives. The mass vaccinations ordered by the Indian government in 1962–1966 proved to be ineffective, even though upwards of 440 million people were vaccinated over that period. Once again, they needed to use the model of search and contain, in a country where not even the chief campaign coordinator had a means of transportation and trips to the most remote areas had to be undertaken by train or bus. But smallpox was gradually eliminated, starting with the wealthier southern states and moving gradually north. Urban slums, where the largest disease epicenters were found, had been contained. When it seemed that smallpox was disappearing from India, civil war erupted in March 1971 to the east of India, in eastern Pakistan (now Bangladesh). The eastern Indian states were literally flooded with Pakistani refugees – an estimated ten million. Fortunately, the Indian medical teams were able to contain most of the refugee camps to prevent the epidemic from spreading. Unfortunately, they were unable to do so in the largest camp, near Calcutta. The smallpox epidemic was unleashed in all its force, but before they could stop it, Bangladesh declared independence. The mass of refugees surged back to their home country, spreading smallpox through eastern India and at home in Bangladesh. Despite enormous efforts, in 1973 there remained four Asian countries where smallpox still existed: India, Pakistan, Bangladesh, and Nepal.

The final part of the campaign began that year, aptly named *Target Zero*. It focused on the three northeastern Indian states that were at the greatest risk of an epidemic. There were 26 special teams, which eventually consisted of 230 epidemiologists from 31 countries, sent into the field. Over 35,000 healthcare workers took part in the campaign, going from village to village and visiting village leaders, teachers, and local businessmen. In time, they visited families. All for the single goal of reporting every case. It was a huge operation, a massive effort. In spite of the effort, in the summer of 1974, less than a year after the intensive campaign was launched, a smallpox epidemic of 116,000 cases broke out in India. The Indian government had to intervene and allot even more money, people, and material to the campaign. It became the government's priority, and the prime minister herself called on the people to give their full cooperation. In June 1974, there were 6,400 infected villages. The houses with smallpox cases had 24-hour guards to make sure the patients did not come into contact with others, and all the people living nearby were vaccinated and quarantined. Food was brought to them so they would not starve to death. In November 1974, the number of infected villages dropped to 340, but in the fall, Bangladesh was hit with some of the worst flooding in history. This resulted in famine, masses of refugees and another outbreak of smallpox. Another 115,000 health workers were enlisted and they went from house to house, searching, isolating and vaccinating. It seemed impossible, but their efforts paid off. The last case of smallpox in India was a 34-year-old beggar living at the train station in Assam, who was identified on May 18, 1975. Three months later, on August 15, 1975, India's Independence Day, Prime Minister Indira Gandhi declared the date as Independence from Smallpox Day. Nepal soon followed, and in October 1975, Bangladesh announced its last case. In January 1976, it was officially announced that smallpox transmission in Asia had ended.

Alongside Asia, problems were also being addressed in Africa, and it was difficult to decide where to begin. Every country was either endemic or shared a border with an endemic neighbor. The problems stacked up: civil strife, famine, refugees from various countries, devastating diseases, authoritarian and often unstable governments, poorly maintained roads, and makeshift bridges.

The work was adventurous, strenuous, and risky. Take for instance what happened to one vaccination team on a helicopter flight. After an emergency landing, they were taken hostage by local rebels and held for a high ransom. After several hours, the ransom was negotiated down to a reasonable amount and the hostages were released. During the negotiations, the rebels had agreed to be vaccinated, so when it was all over the team went out and vaccinated them.

The campaign began in the western part of the continent, which was one of the most poverty-stricken and most infected areas. A total of 20 countries with 116 million potential victims. It continued on to two populous countries in central Africa: Zaire (now the Democratic Republic of the Congo) and Sudan, with a population of 35 million between them. Region three included the countries of east Africa with 50 million inhabitants, and four states and colonies in southern Africa with 45 million. Looking back, it really is incredible that, step by step, country by country, they were able to eradicate smallpox. The disease was eradicated in May 1970 in west Africa. In Zaire it was eradicated in July 1971, with 24 million vaccines administered in this densely populated country over the course of three years. Sudan followed six months later, with the anti-government rebels operating in the southern part of the country playing a role in eradicating smallpox as they smuggled the vaccine for their own people from neighboring Uganda. Smallpox disappeared from east Africa in 1971 and from the southern part of the continent in 1972 and 1973.

All that remained was the Horn of Africa in the east: Ethiopia and Somalia. We now know that the entire campaign ended with the recovery of the last patient, Somali cook Ali Maow Maalin.

Ali was the final chapter in the cultural history of smallpox, one that had lasted over 3,000 years. In the beginning there was the first known case, the Egyptian Pharaoh Ramses V. In the end there was the last known case, a Somali cook. The eradication became one enormous saga in which few at the start had any hope of success. The intensive part of the saga lasted 13 years, from 1967 to 1979, with tens of thousands of healthcare workers around the world participating. It cost nearly $300 million. The United States provided a third of that amount, and two-thirds came from the budgets of other countries.

The campaign was led by Donald A. Henderson, or "D.A." as he was known, who flew thousands of miles around the globe, going to every place where a problem occurred. On one occasion he was almost in a crash, and another time he was nearly shot by insurgents. After completing his mission, Henderson went to Johns Hopkins University, where he became a professor of epidemiology and the Dean of the Johns Hopkins School of Public Health. He died on August 16, 2016.

Henderson deserves much of the credit for the 33rd World Health Assembly's declaration on May 18, 1980 that smallpox had been eradicated from the world. A disease that was the leading cause of death in the 18th century, claiming 400,000 lives annually. A disease that in the 1950s, just before the eradication campaign began, afflicted 50 million Europeans every year and caused two million deaths annually. But no more.

Story 10.8: A gastroenterologist and one of the worst hoaxes in medicine

This story has all the features of a classic drama. It has exposition, plot, climax, twist, and resolution. Everything you need.

We will start with the exposition, the introduction of the story and the main characters. On February 28, 1998, the prestigious British medical journal called *The Lancet* published an article on pages 637–641 (volume 351, issue 9103) by a 13-member team of authors led by Dr. Andrew J. Wakefield. It was a research paper on gastrointestinal disease with the highly technical title of *Ileal-lymphoid-nodular hyperplasia, non-specific colitis, and pervasive development disorder in children*. The authors, who worked at the Royal Free Hospital, examined the relationship between certain gastrointestinal diseases and neuropsychiatric developmental disorders in children. They studied 12 children, aged three to ten, in the hospital's pediatric gastroenterology unit. All the children had gastrointestinal diseases and they all had some form of behavioral disorder, the most frequent being autism (nine cases). The combined MMR vaccine (measles, mumps, rubella) had been administered to eight of the 12 children prior to the onset of symptoms. The authors concluded in the article that "*We have identified associated gastrointestinal disease and developmental regression in a group of previously normal children, which was generally associated in time with possible environmental triggers*." The paper identified the MMR vaccine as the environmental trigger. At a press conference held before the paper was published, Dr. Wakefield, the lead author, said that until the MMR triple vaccine could be eliminated as a potential trigger of developmental complications, he was in favor of single vaccines.

Because it was a small study and the group was not a very representative sample, the press largely ignored it and the message put forward by the authors was forgotten for several years.

Before we get to the second part of the drama, the plot, we need to get to know Dr. Andrew Wakefield and the MMR vaccine a little better.

Andrew Jeremy Wakefield was born in 1956 in Eton, Berkshire, into a family of doctors. He graduated from St Mary's Hospital Medical School in London in 1981. After a three-year fellowship in Canada, he returned to England and went to work at the Royal Free Hospital, where he began to research a possible link between measles and Crohn's disease, an inflammatory disease of the digestive system. Because the measles vaccine is prepared from the weakened virus of the disease itself, Dr. Wakefield also studied the vaccine. In both cases – the live virus that causes the disease and the weakened virus present in the vaccine – he determined a link with Crohn's disease. Although later researchers were unable to confirm his finding, Dr. Wakefield's publications created the first piece of his puzzle: that the measles virus contributed to the development of intestinal disease. Now he just needed the second piece, which was the relationship between Crohn's disease and autism, and the link between even the weakened viruses – Crohn's disease or other intestinal disease – and autism would be completed.

As mentioned earlier, the name of the vaccine in question is MMR, which is an initialism for measles, mumps, and rubella. It was developed by American microbiologist Maurice Ralph Hilleman, one of the most prolific vaccine inventors. He developed a total of 40 vaccines. It is said that his work has saved more human lives than any other scientist of the 20th century. At first, the three vaccines were administered separately, but since the early 1970s it has been a combined vaccine for the simple reason that it is only one injection instead of three, which makes it easier on the children and on their parents, too. The MMR vaccine is used worldwide, with an estimated 500 million doses already administered, and the benefits of the vaccine have been well documented. For measles alone, vaccination in the U.S. prevented 52 million cases and saved 5,200 lives in the first 20 years of use.

Back to the plot of this drama. It starts a few years after the work of AJ Wakefield and his colleagues was published in *The Lancet* and after Dr. Wakefield published another three papers suggesting that immunization by vaccination is not necessarily completely safe. Those next three papers did not present anything new compared to the 1998 paper, but the press launched a campaign the likes of which had never been seen. Questions about vaccination were splashed across the front pages of newspapers around the world. In 2002, the safety of the MMR vaccine – and surprisingly vaccination in general – was the most written about subject ever. A total of 1,257 articles about vaccine safety were published worldwide. Most were written by people who were not trained in the field and they primarily did not even delve into the scientific essence of the issue. Far more frequently, the articles were just contemplations of the medical and political consequences of the author's claims, including the question of whether or not British Prime Minister Tony Blair had had his son Leo vaccinated. The people, especially in Great Britain where the vaccine affair had really escalated, were very confused. In addition, unlike previous medical scares, this one persisted. News and reports about the dangers of vaccines filled the front pages of newspapers for years, and proponents on both sides of the argument made the rounds of the popular television shows. In fact, the debate over vaccine safety rages on. In 2009, every fourth American parent and every fourth British parent believed that the MMR vaccine was linked to autism.

Then came the climax. Fear of vaccination set in. A quarter of Britain's practical doctors began blaming the government for a lack of oversight concerning the safety of the MMR vaccine. In some parts of London, the vaccination rate was only 61 percent and the situation was similar in other areas of Great Britain. That is far fewer vaccinated people than would be needed to prevent an epidemic in any of the given areas – the immunity threshold is around 95 percent – and is one reason the low percentage of vaccinated children is credited for the sudden increase in measles cases in one of those areas: south Wales, where 1,219 children contracted measles. A low vaccination rate is also attributed to the mumps epidemic of 2004. There were 5,000 mumps notifications over the course of the year, as compared to the previous annual notifications that ran into only the tens to the hundreds of cases.

By contrast, there was only a 12 percent overall drop in the UK in willingness to be vaccinated. In 1998, the immunization rate was 92 percent; in 2003, it was down to 80 percent. That is lower than the immunity threshold, but it also indicates that the majority of the population remained unaffected by the huge pressure laid on by the media. The willingness to receive other vaccines actually remained unchanged compared to the period before the panic broke out. Similar results were seen in the U.S., where the number of unvaccinated children only rose from 0.77 percent in 1995 to 2.1 percent in 2000.

And now the plot twist. There are actually two in this drama. The first is the response of medical institutions to the findings presented in AJ Wakefield's papers. Bringing into question such an important medical technology like the vaccine, even if it was only one of many vaccines and more or less only in one paper, necessarily elicited a response from others besides physicians and patients. Medical organizations would have to respond, and they did. Large epidemiological studies were conducted, each of them with a substantially larger number of patients than the 12 children in Dr. Wakefield's study. Reports were issued in the U.S. by the Centers for Disease Control and Prevention, the Institute of Medicine of the United States National Academy of Sciences, and the American Academy of Pediatrics. In Britain, it was the National Health Service. A report was even issued by the *Cochrane Library*, a collection of six massive databases of medicine and healthcare. For its broad report, this independent institution used the results of 64 various clinical studies that incorporated nearly 15 million children. All of these institutions reported the same thing. The *Cochrane Library* stated that "*Exposure to the MMR vaccine was unlikely to be associated with autism, asthma, leukemia, hay fever, type 1 diabetes, gait disturbance, Crohn's disease, demyelinating diseases, bacterial or viral infections.*"

The second twist is what happened with Dr. Wakefield. While the first plot twist was not too surprising, this one was a true reversal. It all began in 2004, when British investigative journalist Brian Deer wrote the first article of his series on Dr. Wakefield and his work.

Deer was no novice at the time, having already specialized in investigating the drug industry and medicine for quite some time. In 1986, long before his first article about AJ Wakefield, he had uncovered that results were being falsified in studies concerning the contraceptive pill. One of his investigations led to the recall of a popular antibiotic, and later the recall of an equally popular painkiller. He was definitely not a rookie reporter.

Deer was the first to discover that Dr. Wakefield had a strong conflict of interest. The conflict originated with the fact that the study of the link between the MMR vaccine and autism was allegedly funded by a group of lawyers preparing a lawsuit against the manufacturers of MMR vaccines. Dr. Wakefield was alleged to have received £435,643, and it did not end with financial support because the lawyers were allegedly involved in selecting some of the patients for the study as well. To top it all off, Dr. Wakefield himself had filed a patent application for a rival MMR vaccine, shortly after his paper raising doubts about the vaccine was published in *The Lancet*. More and more information came to light that increasingly raised questions about his paper, the one that had created such a stir. For example, that the doctor had ignored laboratory data that did not fit with his preplanned findings, and that data fitting his findings were lacking statistical validity. He also altered and incorrectly interpreted other data, all with the goal of creating the impression of a potential link with autism.

The conclusion in the article by Brian Deer that was published seven years later was appalling. Three of the nine children reported with regressive autism did not have autism diagnosed at all. Five of the children had developmental issues before the vaccination, while some of the children did not start having behavioral problems until months after vaccination. In nine cases, unremarkable laboratory test results were changed to support "non-specific colitis."

That was just the tip of the iceberg.

The reactions were immediate. The first to respond was the editor of *The Lancet*, right after the first article was printed in 2004. He announced that had the author's conflict of interest been known, the article would never have been published. That same year, ten of the 12 co-authors issued a statement retracting an interpretation of part of the paper as unsubstantiated and inconsistent with the data obtained. The editors of the *British Medical Journal*, another prestigious British publication, joined in, stating there was clear evidence of falsification of data. None of this was very encouraging for Dr. Wakefield.

But the worst was yet to come. In February 2010, almost exactly six years after it was published, *The Lancet* retracted the paper of AJ Wakefield *et al.* The paper is still available to read on the journal's website, but it is stamped *Retracted* so that everyone knows the findings in the paper are unproven. Other journals followed suit.

On May 24, 2010, the *General Medical Council* (GMC) – the institution that registers physicians in Britain – struck Dr. Andrew J. Wakefield from the medical register. The GMC's ruling stated that Dr. Wakefield "*failed in his duties as a responsible consultant*"; "*acted against the interests of his patients*"; "*acted dishonestly and irresponsibly*"; and "*brought the medical profession into disrepute*." Strong words.

Dr. Wakefield fought back tooth and nail, and in fact he continues to deny any wrongdoing. He denies all the charges and thinks they are an attempt to discredit not just him personally, but also his criticism of the safety of vaccines. He wrote a book in his defense that was published in 2010 under the title *Callous Disregard: Autism and Vaccines – The Truth Behind a Tragedy*. Even this book raised doubts as to the author's relationship with actual facts.

Now to the final part of the drama: the resolution. Dr. Wakefield's work has been called "*the most damaging medical hoax of the last 100 years*," and the doctor himself is noted to be "*one of the most reviled doctors of his generation*." The media did not escape criticism either, because "*the original paper received so much media attention, with such potential to damage public health, that it is hard to find a parallel in the history of medical science*."

The most important thing is that faith in the effectiveness and safety of vaccinations has gradually been restored, almost to the same level as before the first publication of the "*most damaging medical hoax*." Ironically, with the growing doubts about the accuracy and propriety of Andrew J. Wakefield's work, the number of his proponents also grew. There was a polarization of sorts. On one side were respected medical institutions, represented by medical companies and medical journals. They rely on generally accepted procedures for evaluating whether scientific medical information is true or not, from the analysis of laboratory results, to monitoring clinical studies, to collecting and summarizing everything published about the given subject. The main part of this story focused on their opinions, but there is another position as well. While the official position has rejected the work and findings of Dr. Wakefield and has named him "*one of the most reviled doctors of his generation*," the other side thinks of him as "*Nelson Mandela and Jesus Christ rolled up into one*." Supporters of this opinion reject the findings of official institutions and believe they are the result of a conspiracy against not only Dr. Wakefield, but against the people too, a conspiracy that refuses to see the relationship between vaccinations and the rising rate of autism and is therefore causing "*the most catastrophic epidemic of our lifetime*." The United States has the largest number of anti-vaxxers, but there are people all over the world who support the anti-vaccine movement.

How does the story end? Maybe the best way would be to reiterate the conclusion of the *Cochrane Library*, one of the largest studies ever to assess the link between the MMR vaccine and various diseases. We certainly cannot accuse its authors of bias.

"Exposure to the MMR vaccine was unlikely to be associated with autism, asthma, leukemia, hay fever, type 1 diabetes, gait disturbance, Crohn's disease, demyelinating diseases, bacterial or viral infections."

Concluding remarks

One of the many myths circulating about vaccination is the claim that it is a product of modern, scientific medicine. The complete opposite is true: vaccination can actually be considered one of modern medicine's resources. The first methods of immunization were used long before Louis Pasteur and Robert Koch understood the origin of diseases and laid the foundations for today's medicine. Vaccination is also one of the few instances where folk medicine procedures were fully accepted by academic medicine. Just remember the beautiful and wise Lady Mary Montagu's Ottoman method, or how the wise farmer Benjamin Jesty visited the infected cows. Vaccination was also the first method that could fight infectious diseases. Long before Paul Ehrlich, Gerhardt Domagk, or Alexander Fleming discovered their groundbreaking drugs.

The WHO currently lists 27 diseases that can be prevented with vaccination. They are: typhoid, pertussis, poliomyelitis, rotavirus, Hemophilus influenzae type b, Dengue fever, cholera, influenza, Japanese encephalitis, tick-borne encephalitis, malaria, meningococcal meningitis, measles, chicken pox, pneumococcal disease (such as pneumonia or pneumococcal meningitis), human papillomavirus (HPV), rubella, anthrax, tetanus, tuberculosis, mumps, diphtheria, yellow fever, and hepatitis A, B, and E. Several billion vaccine doses are produced and administered every year to prevent these diseases. The WHO estimates that vaccination prevents about 2.5 million deaths annually.

Vaccination helped completely eradicate one disease from our lives and allows us to keep almost all infectious diseases under control in the Western world. However, there are still large areas of the globe where, despite the best efforts of the WHO and many other governmental and volunteer organizations, vaccination does not cover enough of the population. Vaccination in these countries could prevent the deaths of 1.5 million children.

We will end with a comparison offered by one immunologist as his answer to what vaccination is. He said that vaccination can be viewed as the domestication of a wild animal, such as a dog. With a wolf (the wild version of the dog), as with a virus or bacteria, you first eliminate the dangerous attributes of the natural organism so you can get to those that are necessary and beneficial for humans.

11

Conclusion

You have just finished reading the last of the 56 stories about the history of ten medicines. These stories relayed the destinies of nearly 100 heroes. Apart from an abortionist, a wrongly-accused spy, and one enormous quack, they were all people whose contributions not only gave us those ten medicines, but many others too. The history of medicine is replete with amazing discoveries, bold theories, and good decisions, but there are also uncountable dead ends, incorrect hypotheses, and simple human errors. The latter outnumber the former. The history of medicine is rarely linear, it is more of a meandering path, full of twists and turns, along which people throughout history have walked or run toward their goal of eradicating diseases from our lives.

We can happily say that in most cases, they reached their goal, at least as far as Western civilization is concerned. We no longer die of tuberculosis, syphilis, or other infectious diseases. We do not have to worry that a minor scrape might get infected and potentially be fatal. Epidemics of smallpox, measles, typhoid fever, and many other diseases are things of the past, better left to the history books. Women are free to plan their lives without the imperative to conceive children. In fact, even medical students can only look to pictures in textbooks to learn about the greatest contagions in human history. We have been afforded a luxury that people 100 years ago could not even imagine, which is that we do not have to worry that tomorrow we will get sick and die.

Obviously, it is not that simple. New diseases have taken the place of those that we managed to get under control. Infectious diseases have handed off the baton to diseases of affluence. We also have a much longer life expectancy than our ancestors did, so we face problems of old age that they never had to. While the average person born in Europe 100 years ago could only look forward to about 50 years of life, today that number is 80 and higher.

V. Marko, *From Aspirin to Viagra*, Springer Praxis Books,
https://doi.org/10.1007/978-3-030-44286-6_11

This book ends with a sentiment expressed by one of its heroes, Sir Ronald Ross, who received the Nobel Prize for Physiology or Medicine in 1902 and was the protagonist in one of the stories about quinine. It probably best characterizes most of the crusaders against the diseases, to whose memory this book is dedicated. He said:

"*Medical discovery, like all discovery, requires two rather rare qualities – an acute instinct for the right direction and a burning perseverance in following it up.*"

Selected Bibliography

The following list contains only some of the sources used when writing this book. These were main sources, and they can be a useful reference for further information about the subjects.

Chapter 1. ASPIRIN

Jeffreys D. (2005)
Aspirin; The Remarkable Story of a Wonder Drug. Bloomsbury Publ. Plc, London, ISBN 9781582346007

Mann C.C., Plummer M.L. (1991)
The Aspirin Wars; Money, Medicine, and 100 Years of Rampant Competition. Harvard Business School Press, Boston, ISBN 9780394578941

Stone E. (1763)
An Account of the Success of the Bark of the Willow in the Cure of Agues. In a Letter to the Right Honourable George Earl of Macclesfield, President of R. S. from the Rev. Mr. Edmund Stone, of Chipping-Norton in Oxfordshire. Philosophical Transactions (1683–1775), 53: 195–200, http://rstl.royalsocietypublishing.org/content/53/195

Sneader W. (2000)
The Discovery of Aspirin: A Reappraisal. BMJ, Dec. 23, 321/726: 1591–1594

Vaupel E. (2005)
Arthur Eichengrün – Tribute to a Forgotten Chemist, Entrepreneur and German Jew. Angew Chem Int Ed, 44: 3344–3355

V. Marko, *From Aspirin to Viagra*, Springer Praxis Books,
https://doi.org/10.1007/978-3-030-44286-6

Rinsema T.J. (1999)
100 Years of Aspirin. Medical History, 43, 502–507

Cook E.W. (2014)
Legendary Members: Hugo Schweitzer; The Chemist Club, October 2014, http://www.thechemistsclub.com/legendary-members-hugo-schweitzer/

Raschig F. (1918)
Hugo Schweitzer. Angew. Chem., 97: Aufsatzteil, 89–92

Ambruster H.W. (1947)
Treason's peace; German dyes & American dupes. The Beecham Press, New York https://archive.org/stream/treasonspeaceger00ambrrich/treasonspeaceger00ambrrich_djvu.txt.

Smith R.G., Barrie A. (1976)
Aspro – How Family Business Grew Up. Nicholas International Ltd., Melbourne, ISBN 9780903716062

Miner J., Hoffhines A. (2007)
The Discovery of Aspirin's Antithrombotic Effects. Tex Heart Inst J. 34 (2): 179–186 http://www.ncbi.nlm.nih.gov/pmc/articles/PMC1894700/

Souter K. (2011)
An Aspirin a Day; The Wonder Drug That Could Save Your Life, Michael O´Mara Books Ltd., London, ISBN 9781843176329.

Chapter 2. QUININE

Malaria Site, http://www.malariasite.com/

Rocco F. (2003)
Quinine; Malaria and the Quest for a Cure that Changed the World. Harper Collins Publishers, New York, ISBN 9780060959005

Products of the Empire; Cinchona: A Short History. Cambridge University Library, http://www.lib.cam.ac.uk/deptserv/rcs/cinchona.html.

Sneader W. (2005)
Drug Discovery. A History. John Willey & Sons Ltd, Chichester, ISBN 9780471899792

Keeble K.W. (1997)

A cure for the ague: the contribution of Robert Talbor (1642-81), J R Soc Med 90(5): 285–290

Siegel R.E., Poynter F.N.L. (1962)

Robert Talbor, Charles II, and Cinchona. A Contemporary Document. Med Hist 6 (1): 82–85

Pai-Dhungat J.V. (2015)

Caventou, Pelletier & History of Quinine. J Assoc Phys India 63 (3): 58

Wisniak J. (2013)

Chemistry of Resinous Gums, Dyes, Alkaloids, and Active Principles – Contributions of Pelletier and Others in the Nineteenth Century. Indian J Hist Science 48(2): 239–278

Andrews B.G.

Ledger, Charles (1818–1905). Australian Dictionary of Biography, National Centre of Biography, Australian National University, http://adb.anu.edu.au/biography/ledger-charles-4004/text6339

De Kruif P. (1996)

Microbe Hunters. Harvest Book, Harcourt, Inc., New York, ISBN 9780156027779

Ragunathan V., Prasad V. (2015)

Beyond the Call of Duty. Harper Collins Publishers India. ISBN 9789351172644

Chapter 3. VITAMIN C

Carpenter K.J. (1986)
The History of Scurvy & Vitamin C. Cambridge University Press, Cambridge, ISBN 9780521347730

Brown S.A. (2003)
Scurvy. Thomas Dunn Books, New York, Kindley Edition, 2012

Scurvy. (1956) The Naval Revue 26(2): 156-172

Dunn P.M. (1997)
James Lind (1716-94) of Edinburgh and the treatment of scurvy. Arch Dis Childhood 76: F64–F65 http://www.ncbi.nlm.nih.gov/pmc/articles/PMC1720613/pdf/v076p00F64.pdf

Bartholomew M. (2002)
James Lind and scurvy: A revaluation. Journal for Maritime Research 4(1): 1–14 http://www.tandfonline.com/doi/pdf/10.1080/21533369.2002.9668317

Leach R.D. (1980)
Sir Gilbert Blane, Bart, MD FRS (1749-1832). Annals of the Royal College of Surgeons of England 62: 232–239

Johnsom B.C. (1954)
Axel Holst. J Nutr 53: 1–16 http://jn.nutrition.org/content/53/1/1.full.pdf+html

Holst A., Frölich T. (1907)
Experimental Studies Relating to Ship-beri-beri and Scurvy. J Hygiene 7(6): 634–671 http://jn.nutrition.org/content/53/1/1.full.pdf+html

Nobelprize.org (1965)
Albert Szent-Györgyi. Nobel Lectures, Physiology or Medicine, 1922–1941. Elsevier Publishing Company, Amsterdam, http://www.nobelprize.org/nobel_prizes/medicine/laureates/1937/szent-gyorgyi-bio.html

Sterkowicz S. (2007)
Pamieci Profesora Tadeusza Reichsteina (in Polish). Pismo Pomorsko-Kujawskiej Izby Lekarskiej; Okręgowa Rada Lekarska w Toruniu, Meritum, 4 http://www.oil.org.pl/xml/oil/oil67/gazeta/numery/n2007/n200704/n20070405

Chapter 4. INSULIN

Tattersal R. (2009)
Diabetes, The Biography. Oxford University Press, Oxford, Kindle Edition

Zajac J. *et al* (2010)
The Main Events in the History of Diabetes Mellitus. in Poretsky L (ed) *Principles of Diabetes Mellitus*. Springer Science+Business Media, http://friedmanfellows.com/assets/pdfs/elibrary/Principles%20of%20Diabetes%20Mellitus%20-%20Ch1Final.pdf

Bliss M. (2007)
The Discovery of Insulin. The University of Chicago Press, Chicago, ISBN 9780226058993

Von Englerhard D. (ed) (1989)
Diabetes, Its Medical and Cultural History. Heidelberg: Springer-Verlag, Berlin, ISBN 9786342483646

Barnett D.M., Krall L.P. (2005)
The History of Diabetes in Kanh C.R. *et al* (eds) *Joslin´s Diabetes Mellitus, XIVth Edition.* Lippinkott Williams & Wilkins, ISBN 0781727960

MacFarlane I.A.
Matthew Dobson of Liverpool (1735–1784) and the history of diabetes. Practical Diabetes 7(6): 246–248

Barnett D.M. (1998)
Elliott P. Joslin MD: A Centennial portrait. Joslin Diabetes Center Boston

Mazur A. (2011)
Why were "starvation diets" promoted for diabetes in the pre-insulin period? Nutr J 10: 23

Aminoff F.J. (2011)
Brown-Séquard, An Improbable Genius Who Transformed Medicine. Oxford University Press, New York, ISBN 9780199742639

Kahn A.
Regaining Lost Youth: The Controversial and Colorful Beginnings of Hormone Replacement Therapy in Aging. Journals of Gerontology, Series A 60(2): 142–147

Bankston J. (2002)
Frederick Banting and the Discovery of Insulin. Mitchell Lane Publishers, Bear, ISBN 9781584150947

Ferry G. (2014)
Dorothy Hodgkin, A Life. Bloomsbury Reader, London, Kindle Edition

Tabish S.A. (2007)
Is Diabetes Becoming the Biggest Epidemic of the Twenty-first Century?. Int J Health Sci (Qassim) 1(2): V–VIII

Chapter 5. PENICILLIN

Bud R. (2007)
Penicillin, Triumph and Tragedy. Oxford University Press, Oxford, ISBN 9780199254064

Bäumler E. (1984)
Paul Ehrlich, Scientist for Life. Holems & Meier Publishers Ltd., London, ISBN 9780841908376

De Kruif P. (1996)
Microbe Hunters. Harvest Book, Harcourt, Inc., New York. ISBN 9780156594134

Lesch J.E. (2007)
The First Miracle Drugs, How the Sulfa Drugs Transformed Medicine. Oxford University Press, Oxford, ISBN 9780195187755

Brown K. (2013)
Penicillin Man: Alexander Fleming and the Antibiotic Revolution. The History Press, Stroud, Kindle Edition

Meyers M.A. (2007)
Happy Accidents: Serendipity in Modern Medical Breakthroughs. Arcade Publishing, New York, Kindle Edition

Bickel L. (2015)
Florey: The Man Who Made Penicillin. Bloomsbury Publishing, London, Kindle Edition

Tucker A. (2006)
E.S. Anderson. The Guardian, March 22, http://www.theguardian.com/society/2006/mar/22/health.science

Jones J.H. (1993)
Bad Blood. The Tuskegee Syphilis Experiment. The Free Press, New York, ISBN 9780029166765.

Ackermann J. (2003)
Obituary: John Charles Cutler / Pioneer in preventing sexual diseases. Post Gazette, February 12.

Vaz M. (2014)
Ethical blind spots: John Cutler's role in India and Tuskegee. Indian Journal of Medical Ethics 11(3) http://www.issuesinmedicalethics.org/index.php/ijme/article/view/2100/4526.

Chapter 6. THE PILL

Jütte R. (2008)
Contraception, A History. Polity Press, Cambridge, ISBN 9780745632711

Eig J. (2016)
The Birth of the Pill: How Four Pioneers Reinvented Sex and Launched a Revolution. Pan Macmillan, London, Kindle Edition
Morris D. (1970)

The Naked Ape (Slovak Edition). Smena Bratislava

Abbot K. (2012)
Madame Restell: The Abortionist of Fifth Avenue. Smithsonian.com, November 27, http://www.smithsonianmag.com/history/madame-restell-the-abortionist-of-fifth-avenue-145109198/?no-ist

Baker J.H. (2011)
Margaret Sanger, A Life of Passion. Hill and Wang, New York, ISBN 9780809067572.

People & Events: Gregory Pincus (1903–1967), The Pill, American Experience. http://www.pbs.org/wgbh/amex/pill/peopleevents/p_pincus.html

Dr. Pincus, Developer of Birth-Control Pill, Dies, Obituary. The New York Times, 1967, August 23, http://www.nytimes.com/learning/general/onthisday/bday/0409.html.

Berger J. (1984)
John Rock, Developer of the Pill and Authority on Fertility Dies. The New York Times, December 5, http://www.nytimes.com/1984/12/05/obituaries/john-rock-developer-of-the-pill-and-authority-on-fertility-dies.html.

The "Marker Degradation" and Creation of the Mexican Steroid Hormone Industry 1938–1945. American Chemical Society, Sociedad Quimica de Mexico, 1999, https://www.acs.org/content/dam/acsorg/education/whatischemistry/

landmarks/progesteronesynthesis/marker-degradation-creation-of-the-mexican-steroid-industry-by-russell-marker-commemorative-booklet.pdf

Cohen G.S. (2002)
Mexico's Pill Pioneer. Perspectives in Health Magazine 7(1), http://www1.paho.org/English/DPI/Number13_article4_6.htm

Recollections of Life at and after Syntex, George Rosenkranz and Alejandro Zaffaroni take stock of their careers at Syntex Corporation. Chemical Heritage, 2005, 23(2), https://issuu.com/chemheritage/docs/syntex_rosenkranz-zaffaroni/0

Wood G. (2007)
Father of the Pill. The Guardian April 15, https://www.theguardian.com/lifeandstyle/2007/apr/15/healthandwellbeing.features1

Chapter 7. CHLORPROMAZINE

Shorter E. (1997)
A History of Psychiatry: From the Era of the Asylum to the Age of Prozac. John Wiley & Sons, New York, Kindle Edition

Ban T.A. (2007)
Fifty Years Chlorpromazine: a Historical Perspective. Neuropsychiatr Dis Treat 4(3): 495–500

Rosenbloom M. (2002)
Chlorpromazine and the Psychopharmacologic Revolution. JAMA 287(14):1860-1861

Fee E., Brown T.M. (2006)
Freeing the Insane, Am. J. Public Health 96(10): 1743

Chiang H.
An early hope of Psychopharmacology. Bromide treatment in the Turn of the twentieth-century Psychiatry. Historia Medicinae 1(1), Princeton University, Princeton, http://www.medicinae.org/e06

Shorter E. (2009)
Sakel Versus Meduna, Different Strokes, Different Styles of Scientific Discovery. J ECT 25(1): 12–14

Sabbatini R.M.E.
The History of Shock Therapy in Psychiatry. Brain & Mind, http://www.cerebromente.org.br/n04/historia/shock_i.htm

Healy D., Shorter E. (2007)
Shock Therapy: A History of Electroconvulsive Treatment in Mental Illness. Rutgers University Press, London, ISBN 9780813554259

Ugo Cerletti 1877–1963. Am. J. Psychiatry, 1999, 156: 630

Jansson B.
Controversial Psychosurgery Resulted in a Nobel Prize. Nobelprize.org. http://www.nobelprize.org/nobel_prizes/medicine/laureates/1949/moniz-article.htm

Moncrieff J. (2013)
The Bitterest Pills: The Troubling Story of Antipsychotic Drugs. Palgrave Macmillan, New York, ISBN 978113727428

Healy D. (1980)
Pioneers in Psychopharmacology. Inter J Neuropsychopharmacol 1: 191–194

Healy D. (2002)
The Creation of Psychopharmacology Harvard University Press, London, ISBN 9780674015999

Dongier M. (1999)
Heinz E. Lehmann, 1911-1999. J Psychiatry Neurosci 24(4): 362

Bourg J. (2007)
From Revolution to Ethics: May 1968 and Contemporary French Thought. McGill Queens University Press, ISBN 9780773531994

CHAPTER 8. PROZAC

Shorter E. (1997)
A History of Psychiatry: From the Era of the Asylum to the Age of Prozac. John Wiley & Sons, New York, Kindle Edition

Healy D. (2003)
The Antidepressant Era, Harvard University Press, Cambridge, Fourth Printing, ISBN 9780674039582

Wurtzer E. (1994)
Prozac Nation, Young and Depressed in America. Riverhead Books, New York, ISBN 9780704302488

Cahn C. (2006)
Roland Kuhn, 1912–2005. Neuropsychopharmacology 31: 1096

Hinterhuber H. (2005)
Laudatio auf Roland Kuhn. Schweizer Archiv Neurologie Psychiatrie 5:156

Platt M. (2012)
Storming the Gates of Bedlam, How Dr. Nathan Kline Transformed the Treatment of Mental Ilness. DePew Publishing, Dumond, ISBN 9780985730109.

Healy D. (1966)
The Psychopharmacologists, Chapman & Hall, London, ISBN 1860366084

Costa E., Karczmar G.A., Vessel S.E. (1989)
Berhard B. Brodie and the Rise of Chemical Pharmacology. Annual Rev Pharmacol Toxicol 29: 1–22

The Julius Axelrod Papers. Profiles in Science, US National Library of Medicine, Bethesda, USA, https://profiles.nlm.nih.gov/ps/retrieve/Narrative/HH/p-nid/9.

Arvid Carlsson, Biographical. Nobel Prizes and Laureates, Medicine, 2000 http://www.nobelprize.org/nobel_prizes/medicine/laureates/2000/carlsson-bio.html.

Healy D. (2004)
Let Them Eat Prozac, The Unhealthy Relationship Between Pharmaceutical Industry and Depression. New York University Press, New York, ISBN 9780814736975

David Wong
Drug Discovery @ Nature.com http://www.nature.com/drugdisc/nj/articles/nrd1811.html

Kramer P.D. (1993)
Listening to Prozac. Penguin Books, New York, ISBN 9780140266719

Mukherjee S. (2012)
Post-Prozac Nation. New York Times Magazine, April 19

Vann M. (2013)
Are We Still a Prozac Nation? Everyday Health 8(13), http://www.everydayhealth.com/depression/are-we-still-a-prozac-nation.aspx

Chapter 9. VIAGRA

Friedman D.M. (2001)
A Mind of Its Own: A Cultural History of the Penis. Free Press, New York, Kindle Edition

McLaren A. (2007)
Impotence: A Cultural History. The University of Chicago Press, Chicago, Kindle Edition

Loe M. (2004)
The Rise of Viagra: How the Little Blue Pill Changed Sex in America. NYU Press, New York, Kindle Edition

De Kruif P. (1996)
Microbe Hunters. Harvest Book, Harcourt, Inc. New York, ISBN 9780156594134.

Grundhauser E. (2015)
The *True Story of Dr. Voronoff's Plan to Use Monkey Testicles to Make Us Immortal*. Atlas Obscura, October 13, http://www.atlasobscura.com/articles/the-true-story-of-dr-voronoffs-plan-to-use-monkey-testicles-to-make-us-immortal

Schultheiss D., Engel R.M.G. (2003)
Frank Lydston (1858–1923) revisited: androgen therapy by testicular implantation in the early twentieth century. World J Urol 21: 356–363

Brock P. (2008)
Charlatan. America's Most Dangerous Huckster, the Man Who Pursued Him, and the Age of Flimflam Three Rivers Press, New York, Kindle Edition

Klotz L. (2005)
How (not) to communicate new scientific information: a memoir of the famous Brindley lecture. BJU International 96(7): 956-957

Goldstein I. (2012)
The Hour Lecture That Changed Sexual Medicine – The Giles Brindley Injection Story. J Sex Med 9: 337-342

Rogers F.
Sex and Food: The World's Strangest Aphrodisiacs Through Time. Alternet http://www.alternet.org/story/154141/sex_and_food%3A_the_world's_strangest_aphrodisiacs_through_time

Osterloh I. (2015)
How I discovered Viagra. Cosmos April 27, https://cosmosmagazine.com/biology/how-i-discovered-viagra

Leigh M., Lepine M., Joliffe G. (1998)
The Big Viagra Joke Book. Metro Books, London, ISBN 1900512645

Chapter 10. VACCINES

Dobson M. (2009)
Disease (Slovak Edition). Slovart, Bratislava, ISBN 9788080858612

Lady Mary Wortley Montagu. The Montagu Millennium. http://www.montaguemillennium.com/familyresearch/h_1762_mary.htm

Pead P.J. (2003)
Benjamin Jesty: new light in the dawn of vaccination. Lancet 362: 2104–09

Riedel S. (2005)
Edward Jenner and the history of smallpox and vaccination. Proc. (Bayl. Univ. Med. Cent.) 18(1): 21–25

Belongia E.A., Naleway A.L. (2003)
Smallpox Vaccine: The Good, the Bad, and the Ugly. Clin Med Res Apr, 1(2): 87–92.

Franco-Paredes C., Lammoglia L., Santos-Preciado J.I. (2005)
The Spanish Royal Philanthropic Expedition to Bring Smallpox Vaccination to the New World and Asia in the 19th Century. Clin Infect Dis 41(9):1285–1289

Lombard M., Pastoret P-P., Moulin A-M. (2007)
A brief history of vaccines and vaccination Rev Sci Tech Off Int Epiz 26(1): 29–48

De Kruif P. (1996)
Microbe Hunters. Harvest Book, Harcourt, Inc., New York, ISBN 9780156594134.

Brown K. (2013)
Penicillin Man: Alexander Fleming and the Antibiotic Revolution. The History Press, Stroud, Kindle Edition

Colebrook L. (1983)
Almroth Wright – Pioneer in Immunology. Brit Med J Sept. 19: 635-640

Szybalski W. (2003)
The genius of Rudolf Stefan Weigl (1883-1957), a Lvovian microbe hunter and breeder. In Memoriam, in Stoika R. *et al* (eds) International Weigl Conference (Microorganisms in Pathogenesis and their Drug Resistance), Sept 11–14

Henderson D.A. (2009)
Smallpox: The Death of a Disease: The Inside Story of Eradicating a Worldwide Killer. Prometheus Books, New York, Kindle Edition

Langer E. (2016)
D.A. Henderson, 'disease detective' who eradicated smallpox, dies at 87. The Washington Post, August 20

Wakefield A.J. *et al* (1998)
Retracted: Ileal-lymphoid-nodular hyperplasia, non-specific colitis, and pervasive developmental disorder in children, Lancet 351(9103): 637–641

Elliman D.A.C., Bedford H.E. (2001)
MMR vaccine—worries are not justified. Arch Dis Child 85: 271–274

Miller E. (2001)
MMR vaccine—worries are not justified, Commentary. Arch Dis Child 85: 271–274

Index

V. Marko, *From Aspirin to Viagra*, Springer Praxis Books,
https://doi.org/10.1007/978-3-030-44286-6

Printed in the United States
By Bookmasters